Cell Communication in Health and Disease

Cell Communication in Health and Disease

· · ·

READINGS FROM
SCIENTIFIC AMERICAN MAGAZINE

Edited by

Howard Rasmussen
Yale University School of Medicine

W. H. FREEMAN AND COMPANY
New York

Some of the SCIENTIFIC AMERICAN articles in *Cell Communication in Health and Disease* are available as separate Offprints. For a complete list of articles now available as Offprints, write to Product Manager, Marketing Department, W. H. Freeman and Company, 41 Madison Avenue, New York, New York 10010.

Library of Congress Cataloging-in-Publication Data

Cell communication in health and disease: readings from
  Scientific American magazine / edited by Howard
  Rasmussen.
      p. cm.
    Includes bibliographical references and index.
    ISBN 0-7167-2224-0
    1. Cell interaction. 2. Cellular signal transduction.
  3. Second messengers (Biochemistry) 4. Cell
  receptors. I. Rasmussen, Howard, 1925–
  II. Scientific American.
    [DNLM: 1. Cell Communication—collected works.
  QH 604.2 C3923]
  QH604.2.C442 1991
  611′.0181—dc20
  DNLM/DLC
  for Library of Congress   90-15723
                                                      CIP

Printed in the United States of America

1 2 3 4 5 6 7 8 9 0 RRD-C 9 9 8 7 6 5 4 3 2 1

CONTENTS

SECTION III:
DISORDERED CELL COMMUNICATION AND HUMAN DISEASE

Note on cross-references to SCIENTIFIC AMERICAN *articles:* Articles included in this book are referred to by chapter number and title; articles not included in this book but available as Offprints are referred to by title, date of publication, and Offprint number; articles not in this book and not available as Offprints are referred to by title and date of publication.

Preface

The hallmark of multicellular life is specialization. Nearly all the cells of any higher organism have the same organelles, the same metabolic pathways, the same genes and most of the same chemical constituents. Each cell, however, is also unique in expressing some particular set of these components to an enhanced degree in order to provide a specific function of use to the organism. Each cell type developed a particular talent: to secrete a specific product, to contract, to transmit an electrical impulse, to provide metabolic fuel, to eliminate waste products, to regulate temperature, or to exchange oxygen and carbon dioxide with the external environment.

Beginning in the nineteenth century, the systematic study of microscopic anatomy was undertaken using fixed and dye-stained sections of the different tissues and cells in the human body. This systematic study led, in a sense, to the development of an encyclopedia of cell structure. With the introduction of the electron microscope in the twentieth century, the encyclopedia was greatly enlarged. Therefore, our knowledge of structural elements common to all cells is quite extensive, as are those structural elements unique to a particular cell type. However, if each single cell, and each conglomeration of cells—a tissue or organ—is to serve the greater society of cells—the intact organism—then specialization of structure must be accompanied by communication. Each group of cells does not function in isolation, nor does it serve its own selfish ends; it serves the needs of the organism. Hence, if cell specialization is the hallmark of multicellular life, then communication is the very basis of multicellular existence.

Over the past century, but particularly in the past 30 years, a great deal has been learned about the dialogue taking place between cells in a given tissue layer, a given tissue, a given organ, and from organ to organ, as well as that between an organism and its environment. This process of discovery is not yet complete, but it is already clear that the intercellular dialogue is a rich one, of many voices speaking many languages and many dialects. It is also evident that the genes that encode the protein molecules that participate in this dialogue—either as signaling molecules or enzymes regulating the synthesis, breakdown, reception or expression of these signaling molecules—represent a significant portion of the total genes whose functions continue to be expressed in the cells of the adult organism, attesting to the critical importance of intercellular and intracellular communication in multicellular life.

To appreciate the richness and diversity of this

dialogue, it is worthwhile summarizing briefly the historical development of our concepts in this field. By the middle of the nineteenth century microscopic anatomy had provided incontestable evidence that the cells in each organ differed in a structural sense from those in each other organ. For example, it was learned that within the epithelial cells that line the surfaces of the various components of the intestinal tract, those lining the stomach differ from those lining the small intestine, which in turn differ from those lining the large intestine. The implication that was drawn from an appreciation of these structural differences was that functional differences exist between cell types.

The first insights into the communication among cells came from the discovery of the neural innervation of muscle cells in the voluntary nervous system and the innervation of all sorts of cells in internal organs by the nerves of the autonomic nervous system.

The second important insight was provided by the French physiologist Claude Bernard in the middle of the nineteenth century. He proposed that the cells in a multicellular organism were surrounded by and operated within an internal sea — the body's blood plasma and extracellular fluids — the *milieu interieur*. The development of this concept by the American physiologists L. J. Henderson and Walter Cannon led to the elaboration of Bernard's concept. Cells in a multicellular organism operate within a sea whose composition of sodium, hydrogen, potassium, calcium, magnesium, chloride and bicarbonate ions (as well as glucose and a large number of other constituents) are tightly regulated despite large fluctuations in the intake of the essential elements. Cannon introduced the concept of homeostasis to define "the tendency of systems, especially the physiological systems of higher animals, to maintain internal stability, owing to the coordinated response of its parts to any stimulus tending to disturb its normal state." This concept incorporated two principles, that of internal stability and that of the coordinated response of a variety of cells and organs to maintain that internal stability.

The third important insight came from the recognition that specialized tissue, ductless glands, produced chemical messengers. This was achieved by George Oliver and Edward A. Schäfer in England in 1894 when they showed that "extracts of the medullary portion of suprarenal capsules [adrenal glands] produce striking physiologic effects on muscular tissue generally and especially upon the heart and arteries." Within two years the Americans J. J. Abel and A. C. Crawford identified the active principle in these extracts as epinephrine (adrenaline).

Within a year of his pioneering work, Schäfer recognized the full import of his discovery. In an address to the British Medical Association in 1895, he spoke of "My definite subject — the subject of internal secretions — is one of far-reaching interest, although its full importance has only lately come to be recognized. . . . Some secreted materials are not poured out upon an external surface at all but are returned to the blood. These may be termed internal secretions, and they may be and are of no less importance than the better known and more fully studied ordinary or external secretions. The name gland is one which is usually applied to a secretory organ; and to those which have been believed to furnish only internal secretions the name 'ductless glands' has been applied. It is not, however, the ductless glands alone which possess this property of furnishing internal secretions, for it is clear, according to our definition, that this will apply to any organ of the body . . . in that sense every tissue and organ of the body furnishes an internal secretion."

Thus the concept of chemical signaling was introduced to take its place beside neural (electrical) signaling. With the discovery that cell-to-cell communication at most neural synapses involves the release of chemical neurotransmitters that act on the postsynaptic cell to alter its behavior, the century of chemical signaling was born.

Despite Schäfer's insight that all cells send out chemical messengers, the study of chemical signaling during the first half of the twentieth century was largely confined to a study of classic neurotransmitters and hormones manufactured and secreted by identifiable endocrine glands, such as the pituitary, thyroid, adrenal, parathyroid, testes, ovary and endocrine pancreas. The emphasis was on the isolation and characterization of the products of these glands and the definition of their physiologic functions. It was a widely held view that this endocrine system, along with the autonomic nervous system, provided all the signals necessary for the maintenance of homeostasis. The discovery of the hypothalamus as a neuroendocrine organ, which in a sense coupled these two major regulatory systems, seemed to provide the last link in support of such a view. However, in the past 40 years it has become clear that Schäfer's original insight was prophetic. It has been found that nearly all cells send out chemical signals, ranging from derivatives of arachidonic acid, such as prostaglandins, thromboxanes and leukotrienes, to a variety of vasoactive peptides, growth factors and what might be called classic hormones.

In particular, it has become apparent that the different cell types that make up a particular tissue communicate with one another locally. For example, the innermost layer of a blood vessel is made up of endothelial cells and the next layer of smooth muscle cells. Until recently the former were thought to simply perform a barrier function between the blood and the smooth muscle cells. The smooth muscle cells were considered the key cells in regulating blood flow by changing their state of tone or contraction. Muscle tone, in turn, was thought to be regulated either by locally released neurotransmitter, for example, norepinephrine, or by specific peptide hormones, such as angiotensin II or arginine vasopressin. Within the past five years, however, it has become clear that endothelial cells release locally a potent stimulator of contraction, a peptide called endothelin. In addition, these same cells can release a substance that also causes relaxation of the muscle. This is only one example of a common phenomenon — that of local chemical signaling between the various cells in a particular tissue.

In addition, nonclassical endocrine organs such as the heart, liver, kidney and gastrointestinal tract secrete specific hormones. Thus, for example, the kidney secretes erythropoietin, a peptide hormone that regulates red blood cell production by the bone marrow, and 1,25-dihydroxyvitamin D_3, a sterol hormone that acts on the small intestines and bone to mobilize calcium ions into the bloodstream. Likewise, as discussed in Section I, the heart secretes a peptide hormone important in the regulation of blood pressure, the gastrointestinal tract is perhaps the largest endocrine organ in the body and a complex set of hormones regulate the proliferation and differentiation of cells in the bone marrow.

Chapters 1, 2 and 3 emphasize the near-universality of chemical signaling between cells, but only touch on the diversity and commonality of such signaling. A particularly important class of newly discovered substances are tissue-specific growth factors. These are a large and increasing number of peptides, such as nerve growth factor, epidermal growth factor, insulinlike growth factors, platelet-derived growth factor, fibroblast growth factor, and various colony-stimulating factors discussed in Chapter 3, "Hormones that Stimulate the Growth of Blood Cells." In addition, it has been found that many classic hormones, such as angiotensin II, arginine vasopressin and serotonin, can, in addition to regulating tissue or organ function, act as growth factors for certain target cells.

Some of these growth factors regulate the growth and maturation of specific cell types produced by the bone marrow, while others stimulate the growth of various cells involved in wound healing or the growth of cells in any organ that is continually stimulated by one of its tropic factors. In addition, a number of cytokines (a special name for growth factors in the immune system) are the means by which monocytes and lymphocytes in the immune system communicate with one another.

Of equal significance, many peptides that were first identified as hormones are now known to function as neurotransmitters in both the central nervous system and the enteric nervous system. Thus, for example, cholecystokinin, which was first discovered as a peptide released by endocrine cells in the gastrointestinal tract to cause the contraction of the gallbladder, has since been shown to act as a neurotransmitter in the central nervous system at a site that may be important in regulating the response of the organism to an increase in blood glucose concentration and to act as a neurotransmitter in the enteric nervous system to cause the release of insulin from the beta cell.

This example points up two confusing issues about our present views of intercellular communication. One involves nomenclature, and the other involves definition of function.

In the case of nomenclature, a newly discovered chemical signal is often identified either by its source, for example, platelet-derived growth factor, or by its initially discovered effect, for example, somatostatin, a peptide that inhibits the secretion of somatotropin from certain cells in the pituitary gland. Since its discovery, platelet-derived growth factor has been shown to be secreted by cells other than platelets, for example, monocytes, and somatostatin is now known to inhibit insulin and glucagon secretion from the endocrine pancreas and inhibit the secretion of various substances in the enteric endocrine system. Hence, the original name given to a particular extracellular chemical signal may not define either the exclusive tissue of origin or the predominant effect of the particular chemical messengers.

Similar ambiguities have developed in the case of function definition. It was originally clear that acetylcholine, for example, is a neurotransmitter and that insulin is a specific hormone, but it has since become evident that the same substance can serve in a variety of intercellular signaling modes. These range from being an autocrine factor, one that stimulates its cell of origin; a paracrine factor, one that stimulates its neighboring cells in the same tissue; an endocrine factor or classic hormone, one that is released into the blood stream and acts on a target cell at some distant site; neurocrine factor, one that is released as a neurotransmitter but acts on many rather than a single target cell, or as a classic neurotransmitter, one released at a synapse that acts locally on a single postsynaptic cell.

Even though our knowledge of the types and varieties of molecules involved in the dialogue between cells has grown enormously in the past 15 years, nearly every week or so a new extracellular signaling molecule is identified. Hence, our knowledge of extracellular signals, their source and the sites and modes of action is far from complete. One can anticipate a continued expansion of knowledge in this field.

In addition to our expanding knowledge of the nature and sites of action of chemical signaling molecules that provide the basis of the dialogue between the cells, there has been an interest in how these molecules act to regulate the functions of their target cells. Initially this work focused on the possibility that hormones served as co-factors of enzymes. This view developed when it was recognized that various vitamins, dietary trace substances, acted as co-factors for specific types of enzymes. It was considered possible that hormones, internally released trace substances, acted in a similar way. Furthermore, based on the recognized diversity of hormone structure and hormonal effects, it was considered likely that a similar diversity of intracellular signaling pathways would be found. It seemed unlikely, for example, that the ability of epinephrine to stimulate the breakdown of glycogen in the liver to cause the release of glucose into the blood stream shared any common signaling molecules with arginine vasotocin acting on the urinary bladder of the toad to alter the water permeability of the epithelial cells lining the surface of this organ. Unlikely as it seemed, we have since learned that the same intracellular signaling events initiate the response of these two tissues.

The major turning point came in the late 1950's and early 1960's when Earl Sutherland and his colleagues discovered cyclic adenosine 3'5'-mono-phosphate cyclic AMP and cAMP. They showed that in the liver epinephrine did not get into the liver cell and interact with an enzyme but rather interacted with a receptor on the cell surface. The interaction of epinephrine with its receptor activated the enzyme adenylate cyclase, which catalyzes the breakdown of adenosine triphosphate (ATP) into cAMP and pyrophosphate (PP). The cAMP generated at the cell surface was then found to diffuse into the cell and activate the system of enzymes involved in catalyzing the breakdown of glycogen to release glucose. It was, however, the next step that changed the nature of the study of hormone action. Sutherland and others soon showed that the adenylate cyclase–cAMP messenger system was a nearly universal constituent of all mammalian cells and that its activation caused not only liver cells to release glucose, but, for example, adrenal cortical cells to secrete the steroid hormone cortisol, the thyroid cells to secrete thyroid hormone and the toad bladder epithelial cells to change their permeability to water. Out of this work emerged the simple concept that using the same intracellular signaling pathway, a variety of cells could be activated to perform their specific functions.

An additional major insight into the molecular basis of cell signaling was provided by the studies of Edwin Krebs showing that cAMP acted in a specific fashion in all target cells—it activated a new kind of enzyme, a protein kinase. This enzyme was found to catalyze a reaction between ATP and protein leading to the production of ADP and a phosphorylated protein. Phosphorylation of the target protein, by changing the covalent structure of the protein molecule, altered its function. Since the initial discovery of cAMP-dependent protein kinase more than a hundred different protein kinases have been identified. These include a variety of Ca^{2+}-regulated kinases, as discussed in Chapters 4, 5 and 6. These and the cAMP-dependent protein kinase act to transfer the terminal phosphate on ATP to a serine or threonine residue on a particular protein. In addition, there is another group of protein kinases that transfer the phosphate to a tyrosine residue. Many of these tyrosine-specific protein kinases are components of the receptors for specific growth factors.

The question that naturally arose was how could a change in cAMP concentration serve as the intracellular or second messenger in so many different cells expressing such different responses. In other words, how was specificity of response achieved.

The answers to this question were quite simple. First, it was shown that the distribution of receptors for specific extracellular signals are usually restricted to a few target tissues. Thus, for example, the receptor for epinephrine is found on the surface of liver cells, but not on adrenal cortical cells. Conversely, receptors for adrenocorticotropic hormone (ACTH) are found on the adrenal but not liver cells. Hence, no matter how high the epinephrine concentration rises in the blood, there will be no recognition of this fact by the adrenal cell, no rise in cAMP in these cells and, hence, no increase in cortisol secretion. Conversely, no matter how much the ACTH concentration increases, the adenylate cyclase in liver cells will not be activated.

The second feature of specificity relates to the functional capacities of particular differentiated cells. A particular type of receptor for epinephrine, the beta receptor, is linked to the activation of adenylate cyclase in whatever tissue is found. Two of the major tissues are liver and heart. When activated in the liver, the major consequence is glycogen breakdown and glucose release, whereas in the heart the major effects are an increase in heart rate and the contraction force of the heart. The difference in response of these two tissues to the same second messenger depends upon the fact that these specialized tissues have a different set of functional proteins whose activities are changed in response to the same second messenger, that is, the protein substrates for the cAMP-dependent protein kinase are different in different cell types. The heart does possess small amounts of glycogen, and this breaks down in response to epinephrine, but the heart does not possess the key enzyme, glucose-6-phosphatase, so the glucose phosphates, generated as a result of glycogen breakdown, cannot be released into the bloodstream but can only be used as fuel by the heart cells. On the other hand, the liver cell does not possess the contractile system of the heart so that no matter how high the cAMP rises, no contraction is found.

It has become evident from this research on the cAMP messenger system that only a few universal intracellular signaling systems operate to couple extracellular messenger to intracellular response. Various features of these are described in Chapters 4, 5 and 6, with a major focus on calcium ion as the intracellular messenger, one of its major intracellular receptor proteins (calmodulin) and its relationship to inositol lipid turnover and the action of a particular enzyme, protein kinase C. This emphasis

should not be construed to mean that the cAMP messenger system is of only historical importance, and the major universal messenger system is the calcium messenger system. Rather, in a historical sense, the years between 1960 and 1975 were ones in which studies of the cAMP messenger system flowered, whereas much more attention has been focused on the calcium messenger system in the past 15 years. As statements in all these articles attest, there is a nearly universal interaction between these two universal messenger systems in the regulation of specific cellular responses. Thus, for example, regulation of insulin secretion by extracellular messengers involves the activation of both the calcium and cAMP messenger systems, which interact in a synergistic fashion in regulating insulin secretion. Likewise, in many epithelial cells, the proliferative response of these cells involves the simultaneous or sequential exposure to several different growth factors, one of which activates adenylate cyclase. Thus, the concept that most physiologic responses are regulated by a single extracellular messenger acting on one particular intracellular signaling pathway has given way to the view that under physiological conditions multiple extracellular signals interact in a temporal sequence to activate various intracellular signaling pathways to produce the appropriate cellular response. It is this feature of cell communication that underlies the very plastic nature of cellular responses.

A second major line of research that developed out of the original work on the cAMP messenger system was the development of the receptor concept, leading eventually to the recognition that the receptor complex in the cAMP messenger system consists of three components, the receptor itself, a coupling protein that binds and hydrolyzes guanosine trisphosphate (GTP)—a G protein—and the effector molecule or cyclase, the enzyme the catalyzes the conversion of ATP to cAMP. These insights led to the recognition of a family of G proteins, one of which was unique to the retina of the eye. Thus, it became evident, as discussed in Chapter 7 on visual excitation, that the reception of light by the retina displays an operational similarity to reception of a hormonal signal by the liver cell. Furthermore, as discussed in Chapter 4, specific G proteins have been found to couple other hormone receptors to the phospholipase C molecule—the enzyme that catalyzes the hydrolysis of phosphatidylinositol 4,5-bisphosphate to yield inositol 1,4,5-trisphosphate and diacylglycerol—two of the important signaling molecules in the Ca^{2+} messenger system.

With the ability to recognize receptors by biochemical and immunological means, another facet of intracellular signaling became evident. Surface receptors are not a static population of proteins that, once they appear on the cell surface, remain there. Rather, they are continually being internalized and replenished on the surface. In addition, binding of a specific extracellular messenger to its receptor stimulates the process of internalization. If the cells are continually exposed to high concentrations of the specific extracellular messenger, then commonly there is a decrease in the number of receptors on the cell surface—a process known as down regulation of receptor number. This complex process is reviewed in Chapter 7 on how receptors bring proteins and particles into cells. In terms of hormones such as insulin, a major unresolved issue is whether or not the internalized hormone, some of which remains intact inside the cell, serves in any way as an intracellular messenger. On the other hand, as discussed in Chapter 13, the same process of receptor internalization, when the LDL receptor is involved, plays a major role in the regulation of cholesterol metabolism and the development of atherosclerosis.

This new and rapidly expanding knowledge of the nature and actions of the numerous extracellular and intracellular chemical messengers that participate in the dialogue taking place between various cells in man is not only of great biological interest but also of medical interest. As indicated in Chapter 3, the use of erythropoietin in the treatment of certain types of anemias is already being employed, and the potential exists for the use of a number of the colony-stimulating factors for certain other conditions. This use of erythropoietin is similar to hormone replacement treatment in, for example, hypothyroidism, when the thyroid gland fails to make sufficient thyroid hormone to meet the needs of the body.

Such examples represent only a minority of diseases caused by impaired communication among cells, as is made clear in Chapter 9 and is more fully elaborated in Chapters 10, 11, 12 and 13. Even the breadth of these articles fails to indicate the extent of disorders of cell communication that underlie the development of human disease. For example, studies of the mechanisms underlying the growth of normal cells have revealed that nearly every class of cell responds to an appropriate mix of chemical growth factors with a proliferative response. In terms of the intracellular signaling pathways involved, it is common that at least two of the three—cAMP messenger system, Ca^{2+} messenger system and tyrosine kinase-linked pathway—participate in the control of the proliferative response. Not surprisingly, the studies of oncogenes, altered genes which play a role in the abnormal growth of cancer cells, have revealed that many oncogenes are modified components of one or the other of these signaling systems. Hence, the oncogene product may represent the overproduction of a growth factor: an altered receptor, an altered G protein, an altered effector molecule or an altered protein kinase. These discoveries set the stage for understanding the signaling events that underlie normal cell proliferation and for beginning to identify the specific signaling events that become deranged during the development of specific forms of cancer.

Howard Rasmussen

Cell Communication in Health and Disease

THE DIVERSITY OF EXTRACELLULAR MESSENGERS

. . .

Introduction

For a short time after the products of the classic endocrine organs had been isolated and characterized, the belief was held that endocrinology had encompassed the important extracellular messengers involved in homeostasis. However, as the chapters in this section attest, there are a large number of nonclassic endocrine organs that secrete hormones.

Perhaps the most dramatic is the realization that the heart not only pumps blood but stores and secretes a hormone, the atrial natriuretic factor, which plays an important role in the regulation of blood pressure and blood volume, as discussed in Chapter 1, "The Heart as an Endocrine Gland," by Marc Cantin and Jacques Genest.

The extent of the diversity of extracellular messengers and their functions are vividly illustrated in Chapters 2 and 3. Paradoxically, the word "hormone" was first employed by William Bayliss and Ernest Starling in 1905 to describe the properties of secretin—a factor derived from cells in the gastrointestinal tract that simulated pancreatic fluid secretion. Nonetheless, for much of this century the endocrine function of the gastrointestinal tract was considered relatively unimportant. As is made clear in Chapter 2, "The Gastrointestinal Tract in Growth and Reproduction," by Kerstin Uvnäs-Moberg, we now realize that within the various segments of the intestinal tract, there are an enormous number of diverse endocrine cells that secrete 30 or more different substances. These substances have at least three well-defined functions: they integrate the functions of the different cells in a particular segment of the gastrointestinal tract; they help integrate and coordinate the behavior of the different segments of the gastrointestinal tract, and they influence systematic growth and development, particularly in infants and in pregnant women. Hence, there is a very rich and diverse set of chemical messengers arising from the endocrine cells of the gastrointestinal tract that influence not only local events but play crucial roles in nutrition, growth and development.

Chapter 3, "Hormones that Stimulate the Growth of Blood Cells," by David W. Golde and Judith C. Gasson, describes the complex endocrine system, which involves a diverse group of hormones that regulates the selective proliferation in and release from the bone marrow of the platelets, various types of white blood cells and red blood cells. In addition, many of these same hormones act on the mature cells after their release from the bone marrow to regulate their specific functions.

These three examples are representative of a large neoclassical endocrine system. There is growing evidence that nearly every cell type and every organ in the body produces hormones that are released into the blood stream and act on distant cells or tissues to influence their behavior. Many of these hormones remain to be discovered and characterized.

The Heart as an Endocrine Gland

More than pumps, the atria secrete a recently discovered hormone, atrial natriuretic factor, that interacts with other hormones to fine-tune control of blood pressure and volume.

• • •

Marc Cantin and Jacques Genest
February, 1986

The heart is a pump: a muscular organ that contracts in rhythm, impelling the blood first to the lungs for oxygenation and then out into the vascular system to supply oxygen and nutrients to every cell in the body. That has been known since the publication in 1628 of William Harvey's *Essay on the Motion of the Heart and the Blood in Animals.*

Within the past few years it has been discovered that the heart is something more than a pump. It is also an endocrine gland. It secrets a powerful peptide hormone called atrial natriuretic factor (ANF). The hormone has an important role in the regulation of blood pressure and blood volume and in the excretion of water, sodium and potassium. It exerts its effects widely: on the blood vessels themselves, on the kidneys and the adrenal glands and on a large number of regulatory regions in the brain.

The discovery of ANF solved a long-standing mystery. As early as 1935 the late John Peters of the Yale University School of Medicine speculated that there must be a mechanism in or near the heart to "sense the fullness of the bloodstream" and fine-tune the regulation of blood volume. During the 1950's and 1960's numerous investigators searched in vain for a hypothesized "natriuretic hormone."

Such a hormone would explain the occurrence of natriuresis (excretion of sodium) and concomitant diuresis (excretion of water) in the absence of changes in known regulatory processes. Such unexplained natriuresis and diuresis were observed to follow distention of the atria, the two upper chambers of the heart, which receive blood from the pulmonary veins or the vena cava and delivery it to the adjoining ventricles. The putative hormone was referred to as the "third factor," since it would complement the activity of two known regulators of blood pressure and blood volume: the hormone aldosterone and the filtration of blood by the kidney.

The first step toward the discovery of the third factor came in 1956, when Bruno Kisch of the American College of Cardiology noted the presence of what he called dense bodies in the cardiocytes, or heart-muscle cells, of guinea pig atria. In 1964 James D. Jamieson and George E. Palade of the Yale School of Medicine reported that such bodies, whose function was still not known, seemed to be present in the atria of all mammals they examined, including human beings. Our group at the University of Montreal noted in 1974 that the granules were very similar to storage granules seen in the endocrine (hormone-secreting) cells of, for example, the pancreas or the anterior pituitary gland. We found that when radioactively labeled amino acids were introduced into animals, they rapidly appeared in the atrial granules, incorporated into newly synthesized polypeptides (protein chains)— just as they would in the storage granules of endocrine cells.

In 1976 Pierre-Yves Hatt and his colleagues at the University of Paris correlated current knowledge about the granules with earlier findings covering the regulation of sodium and water levels. They showed that the number of granules in the atrial cardiocytes increases when the amount of sodium in an animal diet is reduced (see Figure 1.1). This implied that the granules must store some substance that has to do with sodium balance. A breakthrough was made in 1981, when Adolfo J. de Bold, Harald Sonnenberg and their colleagues at Queen's University at Kingston in Ontario injected homogenized rat atria into rats and observed a rapid, massive and short-lasting diuresis and natriuresis. They concluded that the atria indeed contained a "factor" that promotes these effects, and they named it atrial natriuretic factor.

In the next three years the first direct evidence was reported for the location and biochemical identity of ANF. Workers found there are from two to two and a half times as many of the granules in the right atrium as there are in the left atrium in rats (see Figure 1.2). The granules are highly concentrated near the surface of the heart and in the exterior regions of the atria. They have not been found in the ventricles of rats or any other mammals, and the injection of mammalian ventricular extracts does not affect blood vessels, diuresis or natriuresis. In contrast, granules have been discovered in the ventricles as well as in the atria of nonmammalian species and can be shown to be related to diuretic and natriuretic effects. The presence of the granules in the ventricles of nonmammalian species but not in the ventricles of mammals is consistent with the fact that heart cells tend to be more specialized in higher species.

Once the location of ANF was determined, the peptide was isolated and purified in June, 1983, by our group; it was synthesized two months later by Ruth F. Nutt of the Merck Sharp & Dohme Research Laboratories and her colleagues. ANF is the active part of a larger precursor molecule. When various groups determined the amino acid sequence of the

polypeptide, they all found ANF has the same core of 21 amino acids (see Figure 1.3). The active, circulating hormone in the rat has 28 amino acids and a molecular weight of 3,060. The active hormone is attached to an inactive peptide of 100 amino acids

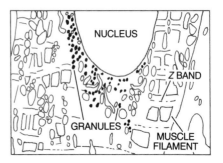

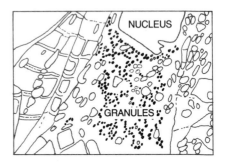

Figure 1.1 HORMONE-STORAGE GRANULES in cardiocytes, or heart-muscle cells, of rats are enlarged some 12,000 diameters in electron micrographs made in the laboratory of M. Cantin and J. Genest. Discovery of such granules first suggested that the heart is an endocrine organ. Stretching of the cardiocyte's contractile apparatus (*filaments with Z bands*) stimulates the release of the hormone atrial natriuretic factor (ANF). In a cell of a normal rate (*top*) granules are seen clustered near the nucleus. In a cell from a rat fed a sodium-deficit diet for 30 days, the number of granules is increased (*bottom*), possibly because the blood volume is lowered. Reduced blood volume decreases the circulating level of ANF, leading eventually to an accumulation of granules in the cell.

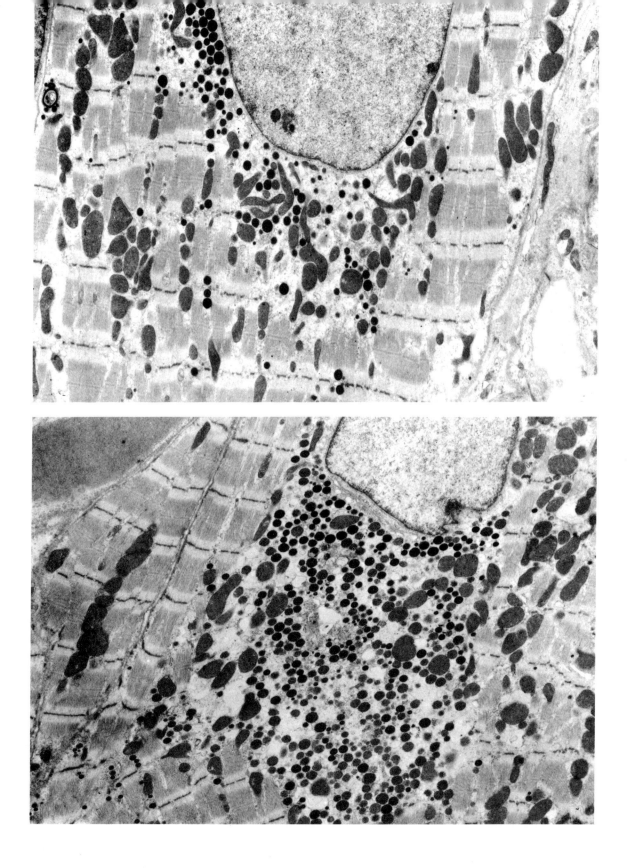

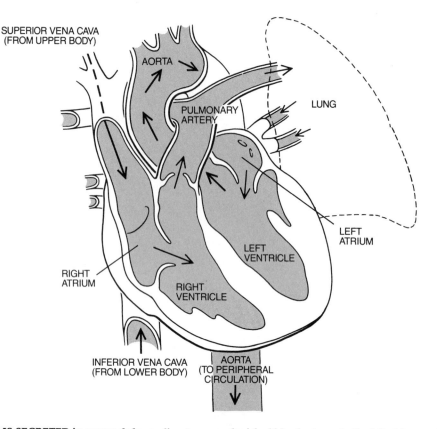

Figure 1.2 ANF IS SECRETED in mammals by cardiocytes in the right and left atria of the heart. Oxygen-depleted blood from the periphery enters the right atrium from the great veins, empties into the right ventricle and is pumped through the pulmonary artery to the lungs. Oxygen-replenished blood returns to the left atrium and is pumped from the left ventricle to the aorta, to be distributed to the periphery. In some nonmammals the ventricles as well as the atria seem to exhibit diuretic and natriuretic activity, and so they too may secrete ANF.

and a 24-amino-acid signal peptide that is cleaved when the molecule is synthesized. The circulating form in the human has not yet been determined, although we strongly suspect that it too is made up of 28 amino acids.

Recently the human ANF gene has been cloned and sequenced, making it possible to synthesize the hormone chemically or by inserting the gene into yeast or bacteria. By either method the hormone can be produced in quantity for studies of its activity throughout the body. In addition antibodies to the hormone have been developed, so that sensitive immunological tests can be carried out to trace the release of ANF and to find the sites where it is active.

When rats are subjected to the stress of immobilization, there is a five- to twentyfold increase in the blood level of ANF. The release of ANF from the heart was also measured in human patients with valvular disease and expansion of blood volume who were undergoing cardiac catheterization, a procedure that allows sampling of blood in the arteries and heart chambers. The plasma level of ANF was from two to eight times as high in venous blood from the coronary sinus, which drains the heart's atria, as it was in blood circulating in the arteries or peripheral veins. This confirms that ANF is released primarily, if not exclusively, by cells in the atria. In both the rats subjected to stress and the catheterized patients the atrial cardiocytes are stretched, and the

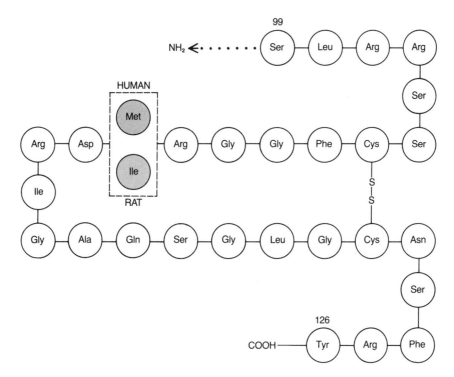

Figure 1.3 AMINO ACID SEQUENCE of the circulating ANF molecule appears to be identical in the human and the rat except at position 110, where the human ANF has methionine and the rat ANF has isoleucine. Both the human and the rat active ANF molecules consist of 28 amino acids, and they both include a disulfide bond between two cysteines that is essential to ANF's activity. The active circulating hormone is cleaved from a much larger precursor polypeptide molecule that is 152 amino acids long in the rat, 151 amino acids in the human.

stretching is the signal for the release of ANF. An increase in blood volume can also cause the atrial cardiocytes to stretch and release ANF, as was shown when the blood volume of rats was experimentally increased by the infusion of a salt solution.

Once the cardiocytes respond to stretching by releasing ANF, the peptide travels through the arteries to targets in the kidneys, the adrenal glands, the brain and various other tissues. In general what ANF does is to modify the activity of a complicated homeostatic feedback loop regulating blood pressure, blood volume and sodium retention: the renin-angiotensin system, which links certain functions of the brain, heart, arteries, adrenals, kidneys and other organs (see Figure 1.4). One of the key substances in the system is the enzyme renin. It is secreted by cells in the arteries that lead to the glomeruli, which are saclike structures at the entrance to the kidney. These juxtaglomerular cells

secrete renin into the bloodstream whenever the level of sodium is low in the distal tubules of the kidney or when the local pressure in the kidney is low.

Circulating renin cleaves a polypeptide called angiotensinogen to produce angiotensin I, which in turn is converted into angiotensin II. This small peptide is a powerful constrictor of vascular smooth muscle. Angiotensin II also has a feedback effect, partially suppressing renin secretion from the juxtaglomerular cells. Finally, it stimulates the adrenal gland to secrete the hormone aldosterone, which travels to the kidney and the posterior pituitary to inhibit the excretion of sodium and water (see Figure 1.5).

ANF affects the renin-angiotensin system by somehow inhibiting the secretion of renin and also by directly inhibiting the adrenal secretion of aldosterone. The relation between ANF and aldosterone was elucidated in our laboratory by a series of ex-

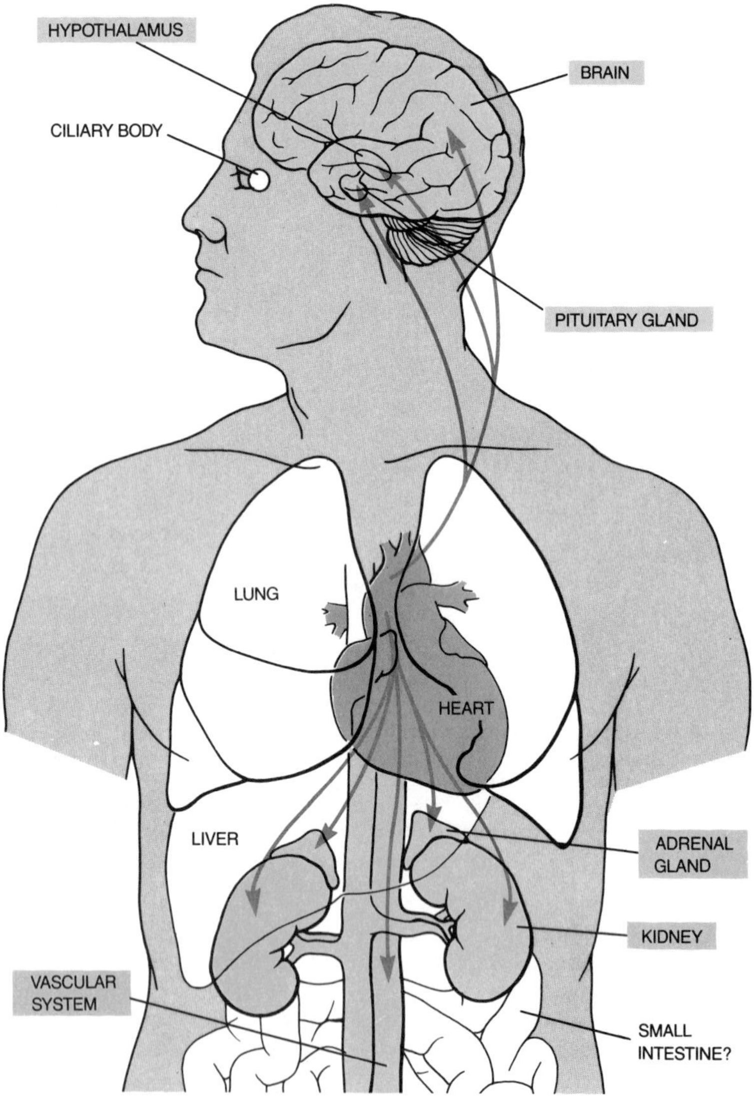

Figure 1.4 ANF IS KNOWN TO AF-FECT various regions of the brain, the posterior pituitary gland, the adrenal gland, the kidney and the vascular system. ANF also binds to targets in the lung, the liver, the ciliary body (which secretes the lymphlike aqueous humor of the eye) and probably the small intestine, but its effects in these tissues are not clear. In the brain ANF binds to various sites that are involved in blood-pressure control and the regulation of sodium and water; in the hypothalamus. ANF inhibits the release of vasopressin, a hormone stored in the posterior pituitary that is antidiuretic and can cause arterioles and capillaries to constrict. ANF relaxes the smooth-muscle cells of blood vessels and inhibits secretion of aldosterone (a hormone that tends to raise blood pressure) from the adrenal gland.

periments with cultured bovine and rat adrenal cells. ANF inhibited the normal production of adrenal aldosterone by 20 percent; it reduced by from 40 to 70 percent the stepped-up production that ordinarily follows stimulation of the adrenal cells with angiotensin II or the pituitary hormone ACTH. Similar significant decreases in plasma aldosterone levels have been noted in rats and dogs after the injection of ANF.

In order to determine whether the decreases in aldosterone levels were caused by the presence of specific binding sites for ANF on the surface of adrenal cells, we introduced radioactively labeled ANF into adrenal-cell cultures and then added "cold," or unlabeled, ANF. We observed that the concentration of the labeled ANF decreased significantly after introduction of the cold ANF, a sign that the cold ANF was displacing the labeled peptide at many sites on the surface of the cells. We determined that ANF binds at very specific sites on the cell by introducing ACTH, angiotensin II and other active peptides. The fact that these peptides did not

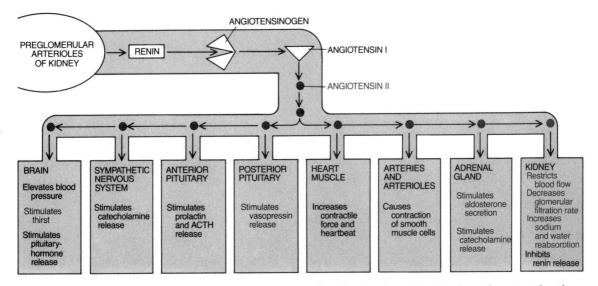

Figure 1.5 ANF INHIBITS certain actions of angiotensin II. When the renal blood pressure is low, the kidney secretes the enzyme renin into the bloodstream. Renin cleaves a polypeptide precursor called angiotensinogen to produce angiotensin I. This is converted in turn into the peptide angiotensin II, which the bloodstream delivers to the organs shown here. Angiotensin II has its hypertensive effects by directly making blood vessels contract, by stimulating the secretion of other contractile hormones (such as vasopressin or adrenaline and other catecholamines) and by stimulating the adrenal gland to release aldosterone. Angiotensin II also has a feedback effect, inhibiting the release of renin from the kidney. Actions known to be inhibited by ANF are shown in color.

exert the same displacement effect as cold ANF testified to the presence of binding sites that are specific for ANF.

In addition to ANF's effects on the renin-angiotensin system, the peptide acts directly at various sites in the kidney to regulate water and sodium excretion. Radioactively labeled samples injected into the rat aorta reveal that ANF binds to epithelial cells in the glomeruli and to numerous receptor sites throughout the blood vessels in the vicinity of the glomeruli and tubules (see Figure 1.6). The hormone somehow exhibits a short-term effect on the mechanism by which the glomeruli filter the blood. ANF probably causes the lining of the glomeruli to become more permeable, allowing larger quantities of water and sodium to be filtered from the blood.

ANF also acts directly in the kidney's tubules, where urine is formed from the filtered blood plasma transported from the glomeruli. The distal tubules "reabsorb" sodium from the filtrate and pass it back into the bloodstream, and ANF probably decreases this activity. The mechanism for ANF's effect in the tubules remains a mystery, since there is no evidence that it requires energy: ANF activity does not require the consumption of oxygen or the breakdown of glucose, unlike other mechanisms of reabsorption.

The role of ANF as a relaxer of muscle cells throughout the vascular system is as important as its functions in the glomeruli and tubules of the kidney. Although injections of ANF can be shown to relax and expand the large vertebral, femoral, common carotid and coronary arteries, the peptide's effect is profoundest in the small arteries of the kidney. When synthetic ANF was added to baths containing strips of rat and rabbit renal (kidney) arteries, the usual constricting activity of angiotensin II and the hormone norepinephrine on the vascular tissue was profoundly inhibited for from 30 to 80 minutes. The infusion of ANF into an isolated rat kidney resulted in a rapid drop in perfusion that lasted for 18 minutes. These findings meant the ANF must be acting either in the smooth-muscle cells of the blood vessels or in the endothelium, the cellular lining of the vessels. When workers added the peptide to blood vessels after destroying the endothelium, the vasorelaxation effect persisted,

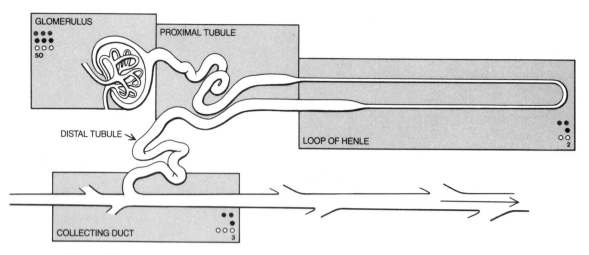

Figure 1.6 IN THE KIDNEY, ANF functions by binding to a target cell, where it often activates particulate guanylate cyclase, an enzyme that synthesises the nucleotide cyclic guanosine monophosphate (cyclic GMP), a second messenger that carries ANF's message to a site within the cell. ANF also inhibits stimulation of adenylate cyclase, which is involved in a different second-messenger system. Studies with radioactively labeled ANF showed a very high density of binding (*colored circles*) in rat glomeruli. Other tests detected marked guanylate cyclase activity (*black circles*) and adenylate cyclase inhibition (*open circles*) in the glomeruli; the cyclic-GMP level (*numbers*) increased more than 50 times over the basal level. The collecting ducts and the thick part of the loops of Henle showed some ANF activity, but the proximal tubules showed none.

indicating that ANF acts on the smooth-muscle cells.

The precise effect of ANF on the smooth-muscle cells is not known. We are doing experiments to test our hypothesis that it affects either the entry of calcium into the cells or its relocation within the cells. ANF could do this indirectly by activating cyclic guanosine monophosphate (cyclic GMP). This is a nucleotide that acts as a second-messenger: a substance that transmits into the interior of the cell the message that has been delivered by a hormone to the cell's surface. Evidence for a relation between ANF and cyclic GMP was demonstrated by the finding that in rats the injection of the peptide leads to significant increases in cyclic GMP in the plasma and urine. ANF may have this effect by activating the enzyme guanylate cyclase, which is attached to the cell membrane and has a role in the activation of cyclic GMP. ANF also inhibits the stimulation of adenylate cyclase, an enzyme that is implicated in second-messenger activity in some cells.

Experiments with radioactively labeled synthetic ANF have shown that the peptide acts at multiple sites in the brain of rats and guinea pigs, binding to areas implicated in the regulation of blood pressure and the control of sodium, potassium and water levels. Investigators have also learned that ANF inhibits the production of vasopressin, a hormone that is synthesized in the hypothalamus at the base of the brain and moves to the posterior pituitary for storage. When vasopressin is released from the pituitary, it constricts blood vessels, raises blood pressure and influences the reabsorption of water by kidney tubules. Finally, experiments have revealed that ANF binds at various sites in the ciliary body of the eye, possibly contributing to the control of ocular pressure there.

While research on the physiology of ANF's effects continues, several groups are investigating how ANF might serve as a drug to control hypertension and congestive heart failure. In our laboratory we have observed a significant short-term (less than one hour) reduction in blood pressure when hypertensive rats are given a single injection of one microgram of synthetic ANF. We found the greatest reductions in animals whose hypertension was dependent on the activity of renin. When ANF was infused at the rate of one microgram per hour for seven days, blood pressure dropped significantly: to normal levels after the second day. A similar treatment with smaller doses administered for 12 days led to a drop in blood pressure over the final 10 days and to lowering of plasma and urinary levels of aldosterone.

In subsequent studies we measured ANF levels in a line of rats genetically predisposed to high blood pressure. We found a high level of ANF in the bloodstream and a lower than normal level in the left (but not in the right) atrium. It would appear that the high circulating level is a sign of the body's attempt to lower the blood pressure; the lower level found in the left atrium suggests that ANF there becomes depleted in hypertension.

ANF appears also to have a major role in congestive heart failure. This is a condition in which the heart fails to pump the blood properly, with the result that the patient becomes short of breath and develops marked edema of the legs. Although it is not known how ANF contributes to the disease, our studies of a line of hamsters afflicted with spontaneous congestive heart failure have correlated changes in ANF levels with the progress of the disease.

In these hamsters the arterial blood pressure is always abnormally low; the venous pressure increases with the severity of the disease. At all stages the amount of ANF in the atria is lower than it is in control animals. The most striking finding is a significant increase in circulating ANF, which is noted as soon as the venous pressure within the heart begins to increase. The level of circulating ANF reaches a peak when the disease is moderately advanced and then decreases in the final stages. Postmortem studies of the diseased atria reveal what we call exhaustion hyperplasia: an increase in the rough endoplasmic reticulum, the site of peptide synthesis; an increase in the size of the Golgi apparatus, where peptides are processed, and a decrease in the number of secretory granules, each of which also harbors less ANF.

These results suggest that a small increase in atrial pressure is enough to trigger hypersecretion of ANF. The decreased ANF level within the atria suggests depletion. We believe sodium retention and activation of the renin-angiotensin-aldosterone system, two indications of congestive heart failure, are postponed until the later stages of the disease by the release of large quantities of ANF. The eventual appearance of these effects may be the result of "down regulation": in the presence of excessive amounts of ANF the target cells may decrease the number of ANF-binding sites on their outer membrane, slowing the cellular reactions that are ordinarily triggered by the binding of ANF.

Further study of ANF should lead to new treatments for hypertension and other blood-pressure diseases, for blood-volume disorders and for kidney diseases affecting the excretion of salts and water. In spite of progress in studying diseases associated with ANF, however, much work lies ahead before synthetic forms of the peptide can be administered to treat patients. The physiology of ANF's effects on the renal tubules remains to be elucidated, as does the relation between the relaxation of blood vessels, calcium movements in the smooth-muscle cells and the effects on adenylate and guanylate cyclases. Workers must also investigate the factors that activate the release of ANF from cardiocytes and determine in detail how the peptide acts in various regions of the brain. Fortunately recent clinical trials have confirmed that all the effects of ANF demonstrated in animals also occur in man; these findings should shorten the route to the development of treatments for disease.

In addition to a more detailed understanding of ANF's physiological activity, progress toward ANF therapy will depend on the development of techniques for tailoring particular analogues of ANF as drugs with which to treat each disorder at a specific binding site. Investigators will have to develop ways of modifying the drugs so that they will be protected from stomach enzymes and acids and can be absorbed easily when they are administered by mouth. Biotechnology and chemical synthesis will provide the needed techniques for these tasks, but several years will probably pass before the first ANF analogues are ready for controlled testing in human patients.

The Gastrointestinal Tract in Growth and Reproduction

It is the largest endocrine gland in the body, and the hormones it secretes exert profound effects not only on the process of digestion, but on the metabolism of ingested nutrients and even the emotions and behavior.

. . .

Kerstin Uvnäs-Moberg
July, 1989

The gastrointestinal tract may seem to be simply a long, winding cavity into which food is introduced, there to be digested into nutrient molecules that can be absorbed through its walls and enter the circulation. In reality the gut is much more than that. It is the largest endocrine gland in the body, and the hormones it secretes exert profound effects not only on the process of digestion itself but also on the metabolism of ingested nutrients and even on the emotions and behavior.

At no stage of life are these physiological functions more critical than during growth and reproduction. Any organism needs more nutriment when it is growing than when it is not. The young of many species eat (relatively) more than adults do; the calorie intake of human infants per kilogram of body weight is more than four times that of adults. The demand for food is also high in organisms undergoing reproduction, which is often preceded by a period of increased uptake and storage of energy. So important is nutrition to reproduction that reproduction simply does not take place in the absence of adequate food.

In mammals, including human beings, most reproductive work is done by the female. Women therefore differ from men in that rather than having only one major period of growth, from infancy through adolescence, they may grow again one or more times as adults: during pregnancy. Beginning early in pregnancy a woman gains weight, storing energy as fat against the demands of the fetus and in preparation for the heavy demands that will come with lactation and breast-feeding. And in human beings, too, energy intake and storage are related to the ability to reproduce. If a woman is too thin, whether because of famine, self-starvation or too much physical exercise, she fails to ovulate and is rendered infertile [see "Fatness and Fertility," by Rose E. Frisch; SCIENTIFIC AMERICAN, March, 1988].

Since increased nutrition is a prerequisite for growth and since food is digested in the gastrointestinal tract, the stomach and intestines need to function optimally during periods of reproduction and intense growth. As the result of a complex reorchestration of the activity of the gastrointestinal hormones, they do just that. In my laboratory at the Karolinska Institute in Stockholm and in many other laboratories, the interacting roles of these hormones have been studied in pregnant women, in the fetus and newborn infant and in both the mother and child during lactation.

To appreciate the role of the gastrointestinal tract in growth and reproduction, one needs first to consider the normal functions of the gut hormones (see Figure 2.1). The hormones are polypeptides: short protein chains of from 10 to 100 amino acids. They are synthesized in specialized endocrine cells whose brushlike projections, the microvilli, project into the stomach and the small intestine. After a meal, different cell types sense either the distension of the wall of the gut or the presence of nutrients or of low pH (acidity) and are thereby triggered to

secrete their hormones both directly into the gut and into the circulation. The activity of the gut's endocrine cells is also influenced by the autonomic nervous system. Activation of the vagal nerve of the parasympathetic nervous system promotes the release of the hormones that enhance digestion; activation of the splanchnic nerve of the sympathetic nervous system has the opposite effect.

The hormones in turn influence gastrointestinal motility and the secretion of digestive enzymes and acid. For example, the hormone gastrin, produced in the lower part of the stomach, enhances digestion by stimulating both gastric motility and secretion of gastric acid. The chemically related hormone cholecystokinin, secreted by cells in the upper part of the small intestine, inhibits the movement of food out of the stomach, thereby enhancing digestion and the absorption of nutrients into the circulation; it also stimulates the release of bile from storage in the gallbladder and the secretion of pancreatic enzymes. The hormone secretin, produced in the same

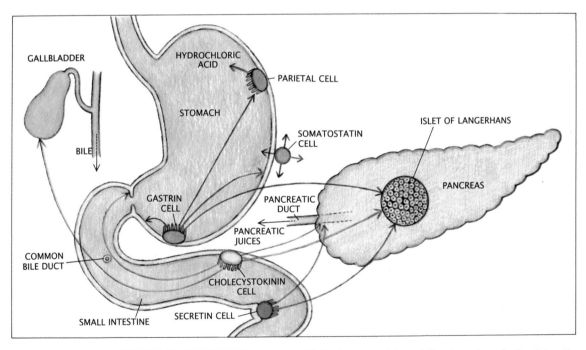

Figure 2.1 GASTROINTESTINAL HORMONES are secreted into the gut and the circulation by endocrine cells in the wall of the stomach and small intestine. Gastrin, cholecystokinin and secretin enhance pancreatic insulin secretion. Gastrin also stimulates release of hydrochloric acid, growth of mucosal cells and gastric motility. Cholecysto-kinin slows emptying of the stomach and stimulates discharge of bile from the gallbladder and secretion of digestive enzymes by the pancreas; secretin stimulates pancreatic bicarbonate secretion. Somatostatin inhibits secretion of gut hormones and counteracts their effects.

area of the intestine, stimulates secretion by the pancreas of bicarbonate (which serves to neutralize gastric acid). Recently it has been established that these polypeptides also stimulate the growth of the organs they affect—in particular a thickening of the mucosa, the organs' mucous lining. In other words, they act in the gut as growth hormones.

One gastrointestinal hormone that does not assume these digestion-enhancing roles is somatostatin. This polypeptide was originally identified in the brain, in the hypothalamus. Now it is known also to be produced in the gastrointestinal tract, where particularly large numbers of somatostatin-producing cells are present in the stomach and in the upper part of the small intestine.

Somatostatin exerts profound inhibitory effects in the gut. For example, it tends to decrease gastrointestinal motility and to block the secretion of hydrochloric acid in the stomach, the discharge of bile from storage in the gallbladder, the absorption of nutrient molecules through the intestinal wall and even the release of such hormones as gastrin and cholecystokinin. Somatostatin also counteracts the growth-promoting effect of gastrin and cholecystokinin in the gut. Because somatostatin is an inhibitory hormone, its neural control is opposite to that of the stimulatory hormones: its release is inhibited by vagal activity and stimulated by splanchnic activity.

Once absorbed, ingested food can be metabolized in essentially two ways. It can go the way of anabolism, the building up from small molecules of larger molecules that can contribute to growth or be stored in the body for future use. Or it can be broken down, in the process called catabolism, to provide energy. When the need for energy is high, as it is during physical exercise and psychological stress, energy is mobilized by catabolism in the liver, muscles and fat tissues in response to the activity of the sympathetic nervous system and the adrenal gland.

In contrast, anabolic metabolism dominates after ingestion of a meal, when fuel for the future is being deposited in the liver and in fat tissues. This storage of nutrients is promoted by the pancreatic hormone insulin, released when blood glucose levels rise in response to digestion of a meal. Eating also increases the activity of the parasympathetic nervous system and thus of the endocrine system of the gut. Several gut hormones, including gastrin, cholecystokinin and secretin, enhance the glucose-induced release of insulin, further stimulating the anabolic type of metabolism. Somatostatin, on the other hand, appears to inhibit the uptake of nutrients and their deposition in storage.

The fact that gut hormones exert an important influence on metabolism is nicely illustrated by some evolutionary data. Sture Falkmer at the Karolinska Institute has shown that the pancreas (which produces insulin) derives from the gut. In primitive vertebrates, which lack a separate pancreas, insulin is produced by endocrine cells in the gut and is released into the circulation after direct contact between those cells and the intestinal contents.

The gut hormones have psychological as well as metabolic effects. In times of stress, which favors the catabolic pathway of metabolism, the level of wakefulness is enhanced. After digestion, on the other hand, when nutrients are being stored, a person often is sleepy and has a sense of well-being. These postfeeding psychic effects originate to some extent in the gastrointestinal tract. Cholecystokinin injected into rats inhibits food intake; Robert S. Mansbach of the University of Vermont in Burlington and his colleagues found that in rats injections of the hormone may even cause electroencephalographic patterns characteristic of sleep. Cutting of the vagal nerve abolishes the inhibitory effect of cholecystokinin on food intake, suggesting that (at least in the rat) information moves from the gut to the brain along neural pathways (see Figure 2.2).

The normal activity of the gastrointestinal tract changes in pregnancy, and the change is an important factor in the characteristic weight gain. Studies conducted in maternity-care centers in Sweden show that women put on an average of 15 kilograms during pregnancy. Part of the gain reflects the weight of the fetus itself, the growth of the uterus and an increase in blood volume, but at least four kilograms represents the deposition of fat.

The simplest explanation for the weight gain of pregnancy would of course be increased food intake. (As folk wisdom has it, "She's eating for two.") Actually, increased intake is only partly responsible. As I mentioned above, women put on weight as early as during the first trimester of pregnancy, when most of them feel sick and consequently eat less, not more (see Figure 2.3). Other physiological mechanisms must therefore be involved, and it is now clear that changes in the hormone activity of the gastrointestinal tract play a leading role.

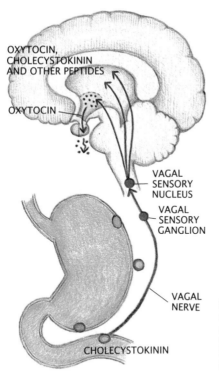

OXYTOCIN, CHOLECYSTOKININ AND OTHER PEPTIDES

OXYTOCIN

VAGAL SENSORY NUCLEUS

VAGAL SENSORY GANGLION

VAGAL NERVE

CHOLECYSTOKININ

Figure 2.2 NEURAL SIGNALS also go from gut to brain. Cholecystokinin in the small intestine triggers vagal sensory impulses; these in turn activate vagal fibers that innervate the brain, promoting maternal behavior, satiety and sleepiness.

My group and others have been able to show that the postprandial release of cholecystokinin increases in pregnant women. By working with pregnant dogs we were able to make repeated measurements that tracked a rise in the average plasma level of cholecystokinin. The elevation is maximal in the first third of pregnancy; from then on it declines gradually but remains clearly elevated above normal until delivery.

The changed hormone levels have several consequences. First, both the increase in cholecystokinin and the decrease in inhibitory somatostatin work to optimize the digestive process. Second, anabolic metabolism is facilitated and weight gain promoted,

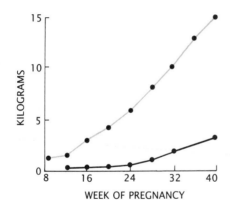

KILOGRAMS

WEEK OF PREGNANCY

Figure 2.3 PREGNANT WOMEN begin to store fat and gain weight (*colored curve*) in the first trimester, when they eat relatively little and before the weight of the fetus (*black curve*) is appreciable. The data are recent averages for Swedish women.

because the levels of the hormones that potentiate the glucose-induced release of insulin rise and the somatostatin level drops. Finally, the enhanced postprandial rise in cholecystokinin is probably responsible—by way of vagal signals to the brain —for the sleepiness and tiredness, particularly after meals, characteristic of early pregnancy. This fatigue has an adaptive effect: it tends to reduce physical activity, so that energy is saved and can be stored. (Of course many working women in advanced industrial societies keep working without letup almost throughout their pregnancy. Physiology suggests that this may actually not be the best course.)

Early pregnancy is characterized not only by fatigue but also by such bothersome symptoms as intense hunger, intermingled with nausea; low blood pressure; and vertigo. The relation of the gastrointestinal tract to such symptoms is clear. Both the frequent hunger and the vertigo may be related to a fall in the blood glucose level, caused in part by the insulin-releasing effect of cholecystokinin and other gut hormones. The sickness and unpleasant sensations in the stomach are likely to be the result of delayed gastric emptying brought about by the high levels of cholecystokinin.

What brings about the changes in the gastrointestinal endocrine system during pregnancy? Hyperactivity of the vagal nerve (which, as I have indicated, modulates gut-hormone release) is likely to be one trigger. Koji Takeuchi of the Kyoto College of Pharmacy in Japan has shown that cutting the vagal nerve of pregnant rats sharply reduces the usual hypersecretion of such digestive juices as hydrochloric acid.

Why should that be? Vagal activity is affected by, among other things, the small neuropeptide oxytocin, which is produced by nuclei in the hypothalamus. Some oxytocin fibers project to the pituitary, from which the peptide is secreted into the circulation; other fibers project to the vagal motor nucleus in the brain stem and stimulate the vagal nerve (see Figure 2.4). Oxytocin secretion is powerfully enhanced by the steroid hormone estrogen, whose level rises in pregnancy. Gut-hormone release may also be affected locally by estrogen and another steroid hormone, progesterone.

We modern human beings have inherited the genetic material and the physiology of our ancestors. For millennia any physiological mechanisms that made more energy available during pregnancy must

have been critically important for the gestation and development of healthy children when food was sparse. In today's advanced industrial societies, characterized (for most people) by an abundance of food, the inherent ability of the female to cut down on energy expenditure during pregnancy is still being expressed—but it tends to be experienced only negatively, as unpleasant pregnancy symptoms and the risk of overweight. In a broader sense, both the ease with which most women put on weight and the higher frequency of obesity in women than in men may reflect women's latent and life-giving ability to store energy, an ability that is fully expressed during periods of active reproduction.

A woman's weight gain in pregnancy is far exceeded by the remarkable rate of growth of infants, who tend to double their birth weight within six months. Their food intake is proportionately high. A six-week-old baby weighing about four kilograms (less than nine pounds) drinks about 650 milliliters (22 ounces) of milk a day; for a 650 kilogram (143-pound) adult the corresponding volume of milk would be 10 liters! The average energy intake of calories in infants is four times as high per kilogram of body weight as that in adults. The gastrointestinal tract needs to be relatively large and highly active in order to cope with the amount of food and calories ingested in infancy. Because most of the nutrients are committed to anabolic metabolism for growth, the endocrine system promoting insulin release can be expected to be particularly active.

In support of this assumption, gastrin levels have been shown (by Alan Lucas of the Medical Research Council Dunn Nutrition Unit in Cambridge and A. Aynsley-Green of the University of Oxford and by my group) to be from five to 10 times higher in newborns than in grown-ups. These high gastrin levels in newborn infants are not secondary to a large intake of food. The baby actually ingests very little milk during the first few days of breast-feeding, and the high gastrin levels precede the high food intake that comes later in infancy.

The very early high gastrin levels can be explained at least in part by the fact that the newborn's gastrointestinal function has been prestimulated during fetal life. Even though nourishment is transferred to the fetus passively, by way of the umbilical cord, the fetal gastrointestinal tract is "trained" in utero for its future task of digesting

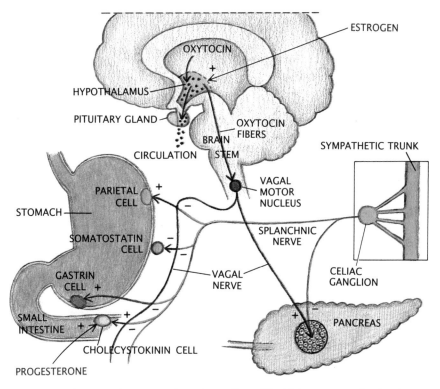

Figure 2.4 NEURAL CONTROL of gut-hormone release is exerted by the vagal and splanchnic nerves. Vagal fibers, activated in part by oxytocin, stimulate (+) or inhibit (−) endocrine-cell hormone secretion and pancreatic insulin release; splanchnic-nerve impulses counteract these vagal effects. The higher estrogen levels of pregnancy stimulate oxytocin secretion, enhancing the vagal effects on gut-hormone and insulin secretion; progesterone seems to act locally to enhance cholecystokinin release.

food. It is known that from time to time the fetus swallows amniotic fluid; swallowing movements have been observed with ultrasound instruments as early as during the first trimester of pregnancy by Heinz F. R. Prechtl of the State University at Groningen in the Netherlands. The amniotic fluid contains several substances, including epidermal growth factor and gastrin, that stimulate gastrointestinal maturation.

The importance of amniotic-fluid ingestion for the development of the fetal gastrointestinal tract was demonstrated by Sean J. Mulvihill of the University of California at Los Angeles School of Medicine. When he ligated the esophagus of fetal rabbits, the development of the gut was severely impaired; normal development was restored when bovine amniotic fluid was introduced into the gut. Ann-Marie Widström, Jan Winberg and I examined the stomach contents of infants immediately after birth. From our data we could conclude that in the human fetus periods of ingestion of amniotic fluid are followed by the secretion of fetal gastrin, somatostatin and gastric acid: apparently even in utero these substances are released in a time pattern resembling what is seen after a meal following birth.

A second reason for the high gastrin levels seen soon after birth may be that the newborn's suckling, by stimulating sensory nerves in the infant's mouth, triggers an activation of the vagal nerves and a consequent release of gastrin and other gut hormones (see Figure 2.5). Giovanna Marchini and I tracked the suckling-related output of gastrointestinal peptides by monitoring their blood levels in plasma. When babies breast-feed, their gastrin, cholecystokinin and insulin levels rise (see Figure 2.6). There is a first peak that is apparently mediated by the vagal nerve: it is notable that the levels rise similarly

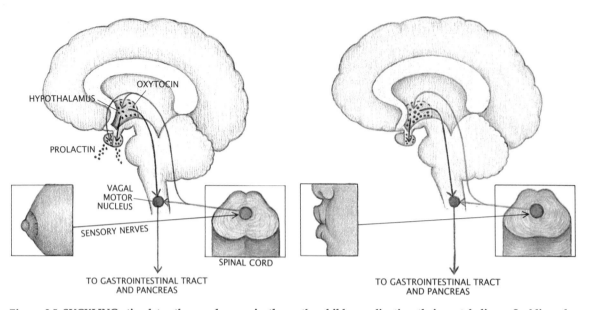

Figure 2.5 SUCKLING stimulates the vagal nerve in the mother (*left*), by way of sensory receptors on the nipple, and in the infant (*right*), by way of receptors in the mouth. Vagal signals alter gut-hormone levels in the mother and the child, coordinating their metabolisms. Suckling also increases maternal prolactin and oxytocin levels, promoting milk production and flow.

when babies merely suck on a pacifier. A second, more protracted hormone response results from the presence of food in the stomach and intestine, but it is probably augmented by the suckling-induced vagal activity.

There is substantial evidence that the suckling process itself has both physiological and psychological effects. Judy C. Bernbaum of the University of Pennsylvania School of Medicine found that infants that need to be fed by way of a nasal catheter grow faster on the same amount of calories if they suck on a pacifier while being tube-fed. Both breast-feeding and sucking on a pacifier sedate a baby and make it sleepy, presumably because suckling increases the cholecystokinin level. Pacifiers, incidentally, are often denigrated as being unhygienic or "spoiling" a child. On the contrary, giving a young infant a pacifier may have real physiological value. After all, the original breast-feeding pattern in humans (which is still seen in some hunter-gatherer tribes) included several periods of suckling per hour, day and night. Perhaps the pacifier serves to compensate for the reduced suckling time that results from the rigid modern pattern.

When infants are sick (whatever the specific diagnosis), they grow slowly and have gastrointestinal symptoms such as gastric retention, constipation and even vomiting. Marchini, Winberg and I found that sick children have 10-fold higher somatostatin levels than do healthy infants. In periods of stress (including stress caused by disease), the sympathetic nervous system is activated, mobilizing energy reserves. At the same time the function of the gastrointestinal tract is inhibited, partly as the result of an increased release of somatostatin. Because somatostatin inhibits not only gastrointestinal motility and secretion but also the release of hormones that stimulate anabolic (energy-storing) metabolism, the peptide is likely to lie behind the retarded growth as well as the impaired gastrointestinal function. Once again there is a clear connection between growth and the gastrointestinal tract—in this case expressed in a negative way.

From a physiological point of view lactation is a continuation of pregnancy: the mother's endocrine and digestive systems continue to provide nourishment for the child. The major differences are that the mother now stores energy in a special depot—the breast—and that the child now re-

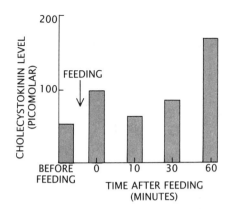

Figure 2.6 CHOLECYSTOKININ LEVEL in an infant is shown before breast-feeding, just after it and 10, 30 and 60 minutes later. The initial rise is a response to suckling; the level rises again when endocrine cells sense the passage of nutrients.

ceives its nourishment in the form of milk rather than from the passage of nutrients through the umbilical cord.

A woman's need for calories is even higher during lactation than in pregnancy; it has been calculated that a lactating woman should increase her intake of calories by 25 percent above her normal intake. How is she to do it? Again the simplest way would be by eating more as the result of an increase in appetite. That does happen. Indeed, it has been shown in rats that the stimulus of the pups' suckling enhances the mother's food intake.

Yet surplus food may not be available during lactation, and so there is a need for physiological mechanisms other than hunger to meet the energy needs of mother and child. One way the mother can decrease her dependence on food is to consume the four kilograms or so of fat stored during pregnancy. That fat tends to be deposited on the thighs and buttocks, and energy is normally rather poorly mobilized from fat at those sites. Per Björntorp of the Sahlgrenska Hospital in Gothenburg has found that its mobilization is facilitated in breast-feeding women by an elegant physiological mechanism: the activity of the enzyme lipoprotein lipase, which promotes fat store, is reduced specifically in thigh and buttock fat during lactation!

The fact remains that not all lactating women have stored fat during pregnancy, and so one can expect there to be other physiological mechanisms that save energy in the mother rather than making more energy available. Recently several investigators have found a discrepancy between the calculated ideal increase in energy intake to support lactation (25 percent in women) and the actual average increase achieved in lactating rats (which lack fat stores) and in women who have not stored fat or who breast-feed for a long time. More specifically, P. J. Illingworth of the Ninewells Hospital and Medical School in Dundee, Scotland, and his collaborators have found that less heat is generated in resting striatal muscles (which normally emit heat after a meal) in lactating women than in nonlactating women. This energy saving may in part explain how lactating women reduce their energy expenditure.

E ven given the energy-mobilization and energy-saving devices, a major factor in supplying the increased energy required for lactation, as for pregnancy, is an increase in the activity of the gastrointestinal endocrine system. The heightened activity is in part secondary to higher food intake but is also caused by the suckling of offspring. Angelica Lindén, Maud Eriksson, Kerstin Svennersten, Widström and I have found, in women as well as in rats, pigs, dogs and cows, that each period of suckling is followed by a release of gastrin, insulin and cholecystokinin and by a decrease in the somatostatin level. The change in the endocrine pattern is a reflex response to suckling, mediated by the vagal nerves: the effects disappear in lactating rats when the vagal nerve is cut. (Indeed, it is likely that the activity of the entire parasympathetic nervous system is increased in response to suckling. The increase would counteract the activity of the sympathetic nervous system, and so it could be the cause of the lessened heat production in muscles, a catabolic process regulated by the sympathetic system.)

The physiological consequences of the release of gut hormones during suckling are significant. First, the digestive process is optimized, and the mucosa

Figure 2.7 NURSING MOTHER AND CHILD are depicted in *Mary and the Child*, a late 15th-century Flemish work. The painting is by the artist known as the Master of the St. Catherine Legend. It now appears that breastfeeding promotes a hormone-mediated physiological symbiosis of mother and child.

of the entire gut is thickened in order to meet the increased metabolic needs of lactation.

Second, energy is stored rather than being consumed for catabolic purposes. Energy is directed to a special depot, the mammary glands, by an increase in the pituitary hormone prolactin, which decreases the number of insulin receptors in maternal fat stores (thus reducing nutrient uptake there) and increases the number of receptors in the mammary glands, leading to an accumulation of nutrients there.

The suckling-related effect on the release of gut hormones contributes to an optimal balance between the intake and the expenditure of energy during lactation, since the amount of milk produced is regulated by the suckling stimulus. Bo Algers of the University for Veterinary Sciences in Skara and I found that suckling by piglets raises the maternal prolactin level and reduces the somatostatin level, and that the extent of these changes varies with the total amount of udder stimulation — its duration multiplied by the number of piglets. Another finding, by Widström and me, further demonstrates the importance of the gastrointestinal endocrine system for the success of lactation: the amount of the decrease in somatostatin level in lactating women that is caused by suckling is highly correlated with the amount of milk taken by the child.

Finally, suckling influences even the behavior of the mother. Women tend to feel sleepy when they are breast-feeding; rats exhibit electroencephalographic patterns characteristic of sleep when they nurse their young. Because cholecystokinin is released from the gastrointestinal tract in response to suckling, the peptide is likely to be involved, by way of the vagal nerves, in these psychological effects. The suckling-induced sedation may serve several purposes. For one thing, it saves energy; it may also help to keep the mother with her offspring.

It is of interest that, as the reader will recall, suckling by the infant triggers a vagally mediated gut-hormone response in the infant as well as in the mother. Indeed, the frequency and intensity of suckling regulate gastrointestinal function in both the mother and the infant and thereby synchronize their metabolisms. In other words, the mother and child are rendered symbiotic not only psychologically but also physiologically (see Figure 2.7).

Hormones that Stimulate the Growth of Blood Cells

*Each hemopoietin regulates the production of a specific
set of blood cells. Now made by recombinant-DNA methods, these
hormones promise to transform the practice of medicine.*

• • •

David W. Golde and Judith C. Gasson
July, 1988

In Leviticus the Bible declares that "the life of the flesh is in the blood." Metaphorical significance aside, the statement is literally true: each type of blood cell is required for life. Erythrocytes (red blood cells) carry oxygen to the tissues. Leukocytes (white blood cells) defend the body from pathogens and tumors. Until recently only a few methods were available to bolster the functions of blood cells: vaccines and better nutrition boosted immune defenses; transfusions might make up for lost red-cell function. Within the last few years a group of hormones have been discovered that may change all that. These protein growth factors are known as hematopoietic hormones or hemopoietins, from the Greek words *haima* (blood) and *poiein* (to make).

Indeed, making blood is just what these hormones do. All the various types of blood cells develop from a single progenitor called a stem cell. Each hormone in the array of hemopoietins causes specific classes of blood cells to be made and "primes" them, enhancing their function. Because genes for several hemopoietins have recently been cloned, these hormones can now be made in quan-

tity, and physicians may soon be able to elicit production of specific types of blood cells as a routine form of therapy. As a result the need for blood transfusions may be greatly diminished, bone-marrow transplantation may be simplified and rendered less risky and the immune system may be bolstered to fight against pathogens, AIDS or tumors. In short, the hemopoietins may effect a revolution in medicine as profound as the introduction of antibiotics half a century ago.

Although red blood cells come in only one form, the white cells of the immune system include three different lineages that carry out specialized functions: granulocytes, monocytes and lymphocytes (see Figure 3.1). The granulocytes are in turn subdivided into three groups called neutrophils, eosinophils and basophils—names that describe the type of stain for which each cell has an affinity. The neutrophil is essential in the host's defense against bacteria and some fungi; the eosinophil has a role in defending against parasites such as worms and protozoans; the function of the basophil is less well

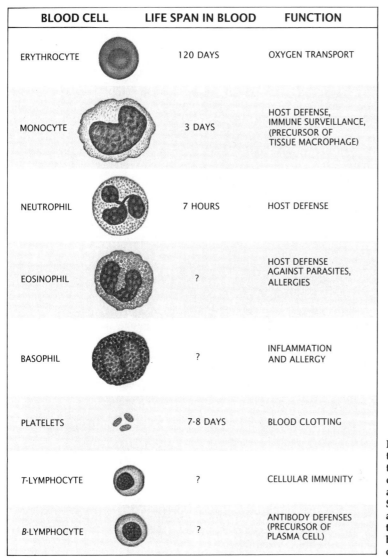

BLOOD CELL	LIFE SPAN IN BLOOD	FUNCTION
ERYTHROCYTE	120 DAYS	OXYGEN TRANSPORT
MONOCYTE	3 DAYS	HOST DEFENSE, IMMUNE SURVEILLANCE, (PRECURSOR OF TISSUE MACROPHAGE)
NEUTROPHIL	7 HOURS	HOST DEFENSE
EOSINOPHIL	?	HOST DEFENSE AGAINST PARASITES, ALLERGIES
BASOPHIL	?	INFLAMMATION AND ALLERGY
PLATELETS	7-8 DAYS	BLOOD CLOTTING
T-LYMPHOCYTE	?	CELLULAR IMMUNITY
B-LYMPHOCYTE	?	ANTIBODY DEFENSES (PRECURSOR OF PLASMA CELL)

Figure 3.1 BLOOD CELLS are of many types, each of which has specific functions. The erythrocyte is the oxygen-carrying red blood cell; all the others are leukocytes (white blood cells). Some monocytes leave the bloodstream and mature into macrophages in the tissues. The neutrophil, eosinophil and basophil are subtypes of the granulocyte.

understood. Monocytes (and related cells called macrophages) are crucial in the defense against intracellular parasites such as viruses and certain bacteria. Lymphocytes help in recognizing and destroying many types of pathogens.

How does this panoply of cell types develop from a single precursor? The process resembles the elaboration of a family tree, with a series of steps taking each descendant farther from the undifferentiated stem cell. Most of this development takes place in the bone marrow, where the stem cells reside. When a stem cell divides, it can replicate itself as a stem cell or become committed to a particular developmental pathway. It is not yet fully understood what governs the "decision" a stem cell makes in becoming committed to a specific lineage. Clearly the process involves expression of some of the cell's genes, but whether commitment is primarily a random process or one dependent on environmental factors is not clear.

In any event, the result of commitment is that the stem cell produces receptors on its surface that respond to specific hormonal signals. Those signals in turn push the cell farther down the pathway toward ultimate specialization. The first major branching of the tree divides the precursors of lymphocytes from those of all other types. At this stage the committed stem cells are not distinguishable on the basis of their form. As differentiation proceeds, the first recognizable precursors appear, including erythroblasts (red-cell precursors) and myeloblasts (precursors of granulocytes and monocytes).

Although it is not difficult to sketch the family tree of the cells of the blood, it is a more complex task to explain how the process is regulated. The pathway leading to mature red blood cells was investigated first and is now fairly well understood (see Figure 3.2). The existence in the blood of a substance regulating the production of red blood cells was postulated early in this century. Since the function of erythrocytes is to transport oxygen, it seemed logical that red-cell production should be tied to the need for oxygen-carrying capacity. Direct evidence for a circulating "factor" that stimulated erythropoiesis (production of red blood cells) in response to changes in the level of oxygen in the environment and in the tissues was provided by Kurt R. Reissmann of the United States Air Force School of Aviation Medicine.

Reissmann connected the vascular systems of two laboratory rats and exposed one partner to low oxygen levels; there was a stimulation of red-cell production in the other partner as well, implying that a blood-borne substance had passed between them. The observations of Reissmann and others paved the way for definitive studies by Allan J. Erslev of Jefferson Medical College. Erslev and his colleagues rendered rabbits anemic by bleeding them. Plasma (the noncellular, liquid part of the blood) from the anemic animals was then injected into normal rabbits, where it stimulated erythropoiesis.

In the years after Erslev's work much was learned about the red-cell growth substance. It was found that the factor is made in the kidney, circulates in the plasma and is excreted in the urine. In laboratory culture its effect is to stimulate incorporation of iron into developing red-blood-cell precursors and to stimulate the growth of colonies of red blood cells. Although much knowledge of the effects of the factor was obtained, the postulated molecule itself was not easy to find—largely because its concentration in body fluids is low. It was not until 1977 that Takaji Miyake and Eugene Goldwasser of the University of Chicago succeeded in isolating

erythropoietin, as the factor came to be called (see Figure 3.3), their efforts yielded only a few milligrams from 2,500 liters of human urine.

Once the protein molecule was in hand, the work moved quickly. Investigators at Amgen and Genetics Institute, Inc., quickly cloned the erythropoietin gene. The purified hormone has a molecular weight of about 34,000 daltons (a hydrogen atom has a weight of one dalton), and the man-made molecule has all the effects observed with the natural one. Those effects have recently been put to use in clinical trials of biosynthetic erythropoietin. Joseph W. Eschbach and John W. Adamson of the University of Washington School of Medicine administered the hormone to patients with severe kidney disease who were dependent on dialysis to free their blood from wastes and toxins.

Dialysis corrects most of the problems caused by severely impaired kidney function except for the accompanying anemia (decrease in the number of circulating erythrocytes). The anemia is due to the low level of erythropoietin, but it may be exacerbated by blood-borne inhibitors and toxins that interfere with the effect of the hormone. The work done by Eschbach and Adamson shows, however, that biosynthetic erythropoietin can overcome the inhibitors, returning the red-cell count to normal. More than 300 dialysis patients have been experimentally treated, and clinically meaningful increases in red-cell production have been seen in almost all of them.

Renal disease is not the only area where erythropoietin will ultimately have profound consequences. Introduction of the hormone will dramatically alter the way blood banks function. Because erythropoietin can theoretically increase red-cell production tenfold, the need for transfusions will decrease greatly. Surgical patients will need fewer transfusions because of erythropoietin's capacity to stimulate their own red-cell production; they may store their own blood and receive erythropoietin before, during and after surgery. In the future the hormone may make it possible for blood banks to grow red cells, thereby transforming the blood bank into a production facility as well as a storage facility. Erythropoietin will also be widely employed to restore red-cell levels in patients suffering from blood-cell cancers or undergoing chemotherapy for many forms of cancer.

Although elucidating the effects of erythropoietin was clearly not a simple job, it was somewhat

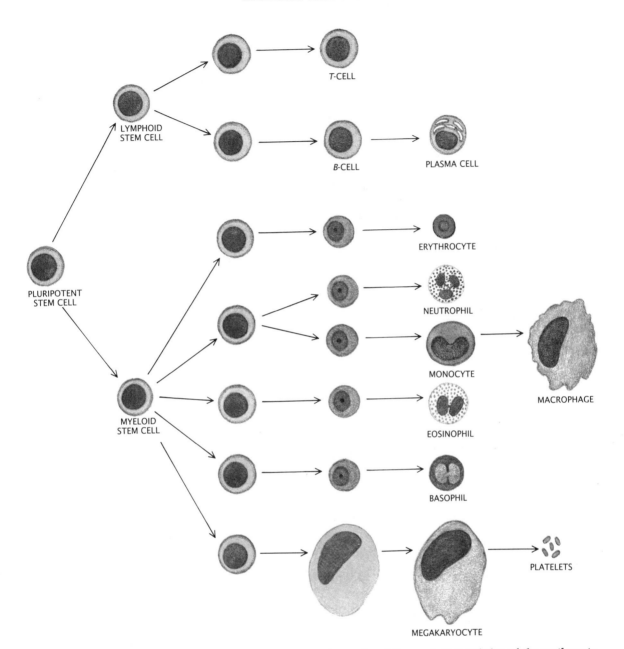

Figure 3.2 BLOOD CELLS MATURE in a pattern resembling a family tree. The pluripotent stem cell, precursor of all mature types, is found in the bone marrow. The first step in the maturation of the blood cells is the division into two main lineages: the lymphoid (consisting of the lymphocytes) and the myeloid (consisting of the erythrocyte and the rest of the leukocytes). Thereafter, under the influence of protein signals, each precursor cell develops step by step into a mature cell type.

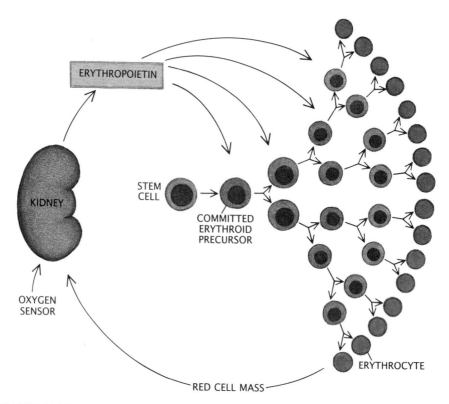

Figure 3.3 ERYTHROPOIETIN is the protein hormone that causes the precursors of red blood cells to proliferate and mature. It is secreted by the kidney in response to the blood's oxygen content. Under the influence of erythropoi- etin, erythrocyte precursors become smaller, make more hemoglobin (the red substance that binds the oxygen molecule) and lose their nuclei as they become specialized for carrying oxygen.

simpler than clearing up the regulatory mechanisms that govern white-blood-cell production. By the early 1960's much work had been done on the rates at which leukocytes develop from their common ancestor. The factors that act on each set of cells, however, were not known. One reason was that until a little more than 20 years ago cells of the human bone marrow could not be grown effectively in laboratory culture. In 1966 that situation was changed by a discovery made independently and at about the same time by Dov H. Pluznik and Leo Sachs at the Weizmann Institute of Science in Israel and by Thomas R. Bradley of the University of Melbourne and Donald Metcalf of the Walter and Eliza Hall Institute of Medical Research in Australia.

What these two teams found was that suspensions containing individual cells from mouse bone marrow could be made to yield colonies of mature white blood cells. Each of the colonies was a clone,

that is, it was made up of genetically identical descendants of a single ancestor. The system's operation depended on the presence in the culture vessel of "feeder" layers to induce and sustain the colonies. At first the feeder layers contained suspensions of various types of cells. Later it would found that the cells were superfluous; it was enough to supply the "conditioned" medium in which the feeder layers has been grown. This was a dramatic step, since it was made clear that the conditioned medium contained substances with the capacity to make white blood cells divide and mature; ultimately such hormones came to be known as colony-stimulating factors, or CSF's (see Figure 3.4).

What were the CSF's and where did they come from? The second question was tackled first. Although most of the early work had been carried out with mouse bone marrow, by 1970 the refine-

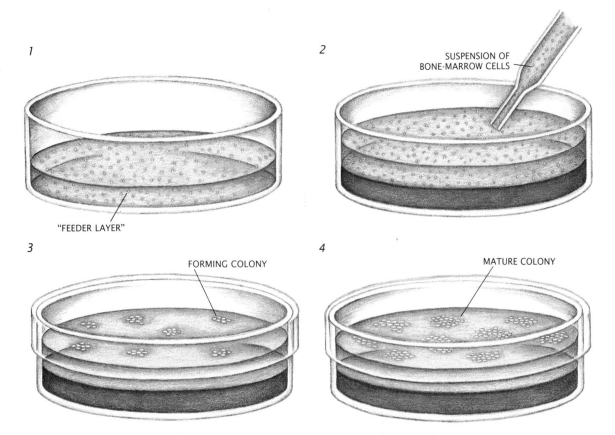

Figure 3.4 COLONY ASSAY led to the identification of the colony-stimulating factors beginning in the late 1960's. "Feeder" layers, containing various types of white blood cells in a semisolid medium, were placed in a small laboratory dish (1). Bone-marrow cells (including stem cells) were added to form a second layer (2). When the dish was incubated, colonies of white blood cells formed in the second layer (3). The colonies were counted and the cells identified (4). When the contents of the first layer were varied, different types of colonies formed, implying the existence of a range of colony-stimulating factors.

ment of the colony-culture system made it possible to grow human leukocytes as a matter of routine. Much of that refinement was due to the work of William Robinson of the University of Colorado. Normal N. Iscove of the University of Toronto and Paul A. Chervenick of the University of Pittsburgh School of Medicine, who developed techniques for growing colonies of human myeloid cells (the lineage that includes erythrocytes, monocytes and granulocytes). Their system entailed feeder layers consisting of human leukocytes or medium conditioned by exposure to the same types of cells.

The finding that human leukocytes released CSF's initiated an intense search for the "colony-stimulat-ing cell" in the peripheral blood that was presumed to be the source of the hormones. The first fruit of that search was the identification of the monocyte as the blood leukocyte primarily responsible for the release of the CSF's; that finding was made by Chervenick and Al F. LoBuglio of Ohio State University and independently by Martin J. Cline of the University of California School of Medicine at San Francisco and one of us (Golde). Later we found that the macrophage (a descendant of the monocyte that is present in the tissues rather than in the blood) also releases CSF's (see Figure 3.5). Even more recently it has been found that cells of the monocyte-macrophage lineage make substances that can cause the

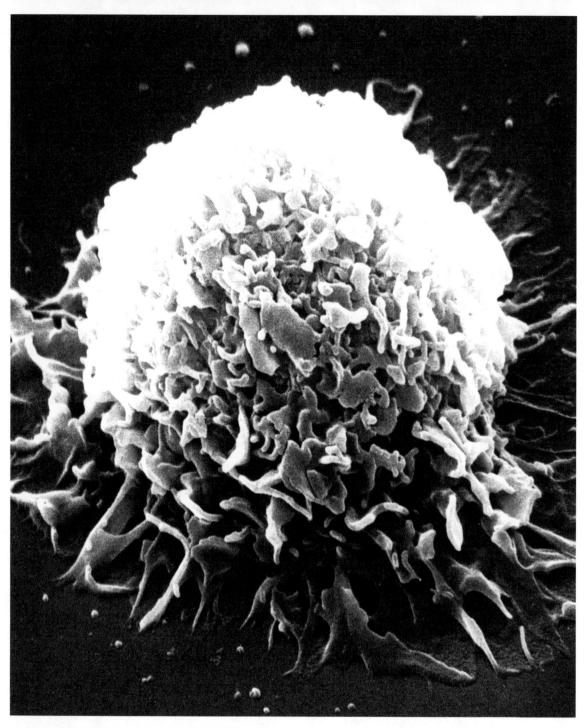

Figure 3.5 MACROPHAGE is a white blood cell that has a central role in the immune response. A motile cell found in many body tissues, the macrophage phagocytoses (ingests) both pathogens and wastes. The "ruffles" on the surface of the cell serve in locomotion, cell spreading and phagocytosis. The macrophage is among the cells that release hormones called colony-stimulating factors (CSF's), which induce white blood cells to proliferate and mature. The macrophage is shown enlarged some 9,000 diameters in this micrograph made by S. G. Quan in the laboratory of D. W. Golde and J. C. Gasson.

release of CSF's from other cells; examples of these substances are interleukin-1 (IL-1) and tumor necrosis factor (TNF). Such findings have led to the idea that the macrophage is a sentinel, responding to microbial invasion and sending signals to trigger increased synthesis of white blood cells.

The monocyte and the macrophage are not the only white blood cells that release CSF's. Further work showed that when lymphocytes are activated, they also release a potent CSF. Lymphocytes come in at least two forms, known as T and B lymphocytes, each with a variety of functions. Some B lymphocytes ultimately differentiate into plasma cells, which make large quantities of antibodies. T lymphocytes have a wide range of roles, including serving as a kind of "master" cell, precisely orchestrating many aspects of the immune response. It is in this capacity that the T lymphocytes make and give off colony-stimulating factors. When they are exposed to specific antigens, they release CSF's that marshal the production of white blood cells.

As the 1970's passed, some of the sources of the colony-stimulating factors were becoming clear. Yet the question of what these molecules actually were had not been answered. The first substantial evidence came from the mouse-bone-marrow colony-assay system described above. In that system four different CSF's were identified. The first, identified by E. Richard Stanley of the Walter and Eliza Hall Institute and Metcalf, only induced colonies of the monocyte-macrophage lineage, and so it was named macrophage CSF (M-CSF). Three other CSF's were identified after that. Of those, the one that stimulated granulocyte and monocyte colonies was called GM-CSF, and the one that stimulated only granulocyte colonies, G-CSF. The third, which resulted in colonies containing a mixture of cell types, was called multi-CSF, or interleukin-3 (IL-3).

Having identified a set of CSF's by their activity in culture, investigators wanted very much to be able to obtain them in pure form. One advantage of having the purified protein in hand was that it could lead to cloning the gene for the protein. Once cloned, the gene could be inserted into mammalian or bacterial cells. There the protein could be made in quantities large enough for research purposes and for therapeutic trials. As it happened, macrophage CSF was the first to be obtained in highly purified form (by Metcalf's group), but it was not the first CSF whose gene was cloned.

That honor went to the murine (mouse) form of GM-CSF. Antony Burgess of the Walter and Eliza Hall Institute purified the protein from medium conditioned by lung tissue from mice that had been injected with endotoxin (a substance found in the cell walls of certain bacteria that triggers a potent immune response). Nicholas M. Gough and Ashley R. Dunn of the Walter and Eliza Hall Institute then employed a partial amino acid sequence of the protein to construct complementary DNA (cDNA) probes. The probes in turn were exploited to pick the GM-CSF gene out of a "library" of mouse DNA sequences. Work on the human hormone lagged, but ultimately our group purified it from medium conditioned by a line of human T lymphocytes that had been transformed by HTLV-II (a human virus that can cause leukemia). Gordon G. Wong and Steven C. Clark of Genetics Institute, collaborating with us, then exploited a novel laboratory system to retrieve the GM-CSF gene from a line of monkey cells.

Thus GM-CSF was the first human hematopoietic hormone whose gene was molecularly cloned; it was also the first to be made by recombinant-DNA methods. Yet others were not far behind. Yu-Chang Yang of Genetics Institute and Clark cloned sequences encoding IL-3; Karl Welte and Malcolm A. S. Moore of the Sloan-Kettering Memorial Cancer Research Institute with Lawrence M. Souza of Amgen (and at the same time Shigekazu Nagata of the University of Tokyo and his colleagues) did the same for G-CSF. While that work was going on, Ernest S. Kawasaki and David F. Mark of the Cetus Corporation cloned part of the DNA for the macrophage CSF. Once the genes had been cloned and the hormones had been made by recombinant methods, the biosynthetic hormones could be tested for their capacity to grow white blood cells. Each of the recombinant hormones has shown the specific activity expected of it on the basis of observations from the colony-assay system.

Although cloning the genes for the human CSF's was an important step, it did not provide direct information about the function in the human body either of the genes or of the corresponding hormones. Using the cloned DNA as probes, however, the human genes that encode the hormones have now been located on specific human chromosomes. Each gene is apparently present in a single copy. (Genes may be present in multiple copies; the purpose of such amplification may be to enable the cell to make more of the gene's product.) The genes for GM-CSF and IL-3 are quite close to each other on chromosome 5, which suggests they may have had a common ancestor. Interestingly, the genes for M-CSF (a protein unrelated to the other two) and its receptor are also on chromosome 5. The G-CSF gene, however, is on chromosome 17.

When are these genes turned on or off in hematopoiesis? Work on that question is proceeding rapidly and is beginning to yield an outline of how the process as a whole is regulated. Some of the CSF genes are not expressed under ordinary conditions; they remain quiescent until the cell receives

a specific signal (see Figure 3.6). GM-CSF, for example, is released by lymphocytes when they are activated by specific antigens (foreign proteins). It is also made and released by fibroblasts (connective-tissue cells that have a role in wound healing) and endothelial cells (cells that line the blood vessels) when these cells are exposed to substances from monocytes and macrophages, such as tumor necrosis factor and IL-1. Macrophages can also synthesize and release GM-CSF when stimulated.

Similarly, the G-CSF gene can be turned on in a

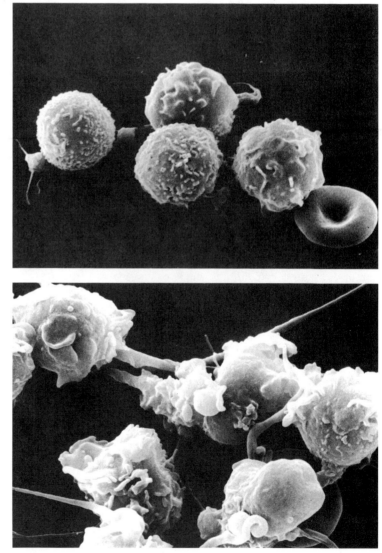

Figure 3.6 GRANULOCYTES ARE ACTIVATED by granulocyte-macrophage colony-stimulating factor (GM-CSF). In the upper panel the granulocytes are in the resting phase (the disklike object at the lower right is an erythrocyte). In the lower panel the granulocytes have been stimulated by GM-CSF. The long projections—called filopodia—enable the granulocyte to adhere to surfaces and to make contact with pathogens. The cells are enlarged some 3,000 diameters in these micrographs by G. E. Garner and L. Hancock, Jr.

number of cell types in the presence of activating signals. It is made by cells of the monocyte-macrophage lineage when they are exposed to endotoxin and by fibroblasts in response to substances released by monocytes and macrophages. M-CSF is also made by many cells, including the macrophage itself, in response to endotoxin, GM-CSF or IL-3. The IL-3 gene is turned on in activated lymphocytes. Thus, in response to a network of intercellular signals, genes for the CSF's are aroused from their quiescent state and begin to give rise to their products.

This network of interactions may appear complex. It is. What is more, the network is not fully understood. What is already known, however, provides the basis for a preliminary outline of how the immune system responds to a pathogen in terms of CSF's. In that picture the T lymphocyte and the macrophage have a central role. In response to a particular antigen a subset of T cells are activated, releasing GM-CSF and IL-3 (see Figure 3.7). At the same time macrophages responding to their specific stimuli synthesize GM-CSF, G-CSF and M-CSF. The release of IL-1 and TNT from the macrophages triggers production of GM-CSF, G-CSF and M-CSF in local populations of endothelial and mesenchymal cells (a type found in muscle, bone and connective tissue). As the result of the release of this array of CSF's (see Figure 3.8), the subpopulations of leukocytes that are required in the immune response begin to proliferate and mature.

Yet the role of the CSF's in immunity does not end there. Although these factors were originally identified by their capacity to stimulate the growth and maturation of stem cells and precursor cells, it seems they also have profound effects on mature white blood cells. Neutrophils activated by invading microorganisms synthesize and release highly toxic oxygen derivatives that can kill the invaders. Richard H. Weisbart of the University of California at Los Angeles (in collaboration with our group) showed that GM-CSF primes the neutrophil, thereby making the response more potent. GM-CSF did not trigger the oxidative burst directly, but it did lead to a markedly increased release of oxidants when the neutrophils were exposed to known triggering agents such as bacterial proteins.

The neutrophil's mission is to "search and destroy": if its weapons are to have an effect on the invader, the enemy must first be found. GM-CSF and G-CSE both increase the directed movement of neutrophils toward triggering agents. The CSF's also augment a neutrophil's capacity to ingest microorganisms. How this happens is not fully understood. Studies by Weisbart and his co-workers, however, have shown that GM-CSE regulates the number and affinity of cell-surface receptors on the neutrophil that recognize bacterial products. Thus as the granulocytes survey the environment within the body, CSF's may render them better able to detect pathogens.

The capacity of the neutrophil for searching and destroying is central to the "host defense" mounted by the immune system. In the end the host defense rests on the ability of mature effector cells—such as neutrophils and macrophages—to kill invading organisms (see Figure 3.9). When the bone marrow cannot make enough neutrophils or monocytes, host defense is impaired. The response of the effector cells also depends on the collaboration of T lymphocytes. When there are too few T cells or the T cells function poorly (as in AIDS or advanced tuberculosis), the macrophage is incapable of containing the intracellular pathogen. Such deficiencies are clearly catastrophic from the clinical point of view; the CSF's, which can help to correct them, have great therapeutic potential.

Some of that potential is already being realized. Studies of several CSF's made by recombinant methods have been carried out in laboratory animals. Such trials indicate that GM-CSF and G-CSF are relatively nontoxic and are effective in stimulating the proliferation of host-defense cells. Human G-CSF causes increased neutrophil production in both mice and monkeys. In monkeys the increase was to a level 50 times the normal one, and there were few side effects. Human GM-CSF is not effective in mice, but in monkeys it yields increases in neutrophils, eosinophils and monocytes. The administration of IL-3 alone to monkeys results in only a modest increase in circulating white cells, but in combination with GM-CSF it has a potent stimulating effect on bone marrow and leukocyte production.

There have so far been only a few trials in human beings of recombinant CSF's, but those that have been carried out tend to confirm the exciting prospects offered by the animal work. Ronald Mitsuyasu, working with us and with Jerome Groopman of New England Deaconess Hospital, conducted the first human phase I and II (safety and efficacy) trials of GM-CSF in patients suffering from AIDS, with its characteristic decrease in white-blood-cell count. This initial study showed that GM-CSF was well tolerated and that it increased levels of neutro-

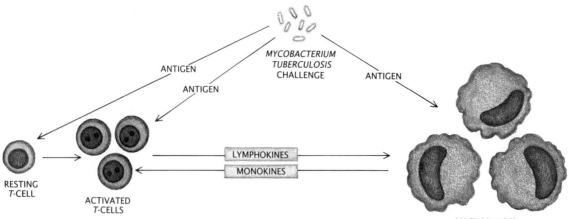

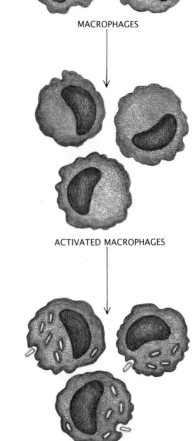

MACROPHAGES

ACTIVATED MACROPHAGES

M-TB INGESTED AND KILLED

Figure 3.7 MACROPHAGES ARE ACTIVATED in a process that entails interacting with *T* lymphocytes. When *Mycobacterium tuberculosis* (the tuberculosis agent) enters the body, its proteins activate a specific subset of *T* cells. Those *T* cells release lymphokines that prime macrophages to ingest and kill bacteria. Among the lymphokines are GM-CSF, interleukin 3 (IL-3) and interferon. Macrophages in turn release monokines, including interleukin 1 (IL-1), which stimulate the lymphocytes.

phils, monocytes and eosinophils circulating in the blood. GM-CSF is also being investigated as a form of therapy in cancer, preleukemic conditions and aplastic anemia (anemia due to malfunction of the bone marrow).

Studies of the other CSF's are not as advanced as those of GM-CSF, but they are gaining momentum. Janice Gabrilove and her colleagues at Sloan-Kettering carried out a study of G-CSF in bladder-cancer patients being treated by combination chemotherapy, which often leads to suppression of bone-marrow function and reduced white-cell counts. The hormone had few side effects, and it clearly stimulated production of neutrophils in the bone marrow and reduced bone-marrow suppression. M-CSF also appears to be relatively nontoxic in monkeys and human beings, but it has not undergone rigorous clinical trials. IL-3 has not been assessed in clinical trials.

The clinical results available so far suggest that the CSF's will one day prove their worth in treating AIDS patients and in overcoming the leukopenia (low white-cell counts) associated with cancer chemotherapy. If one accepts the capacity of these factors to increase host-defense cell numbers and activity,

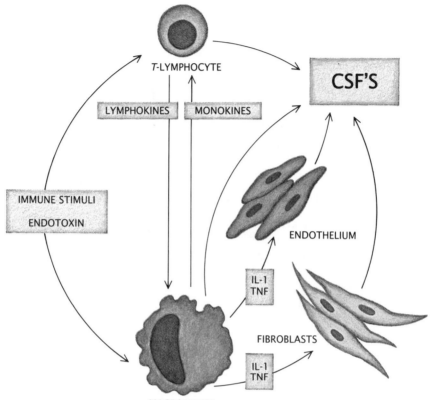

Figure 3.8 COLONY-STIMULATING FACTORS ARE RELEASED by an array of cells in response to the presence of pathogens. Immune stimuli (antigens) cause *T* cells to release CSF's; endotoxins (molecules found in the cell wall of certain bacteria) cause macrophages to do the same. In addition, the macrophage releases substances that evoke release of CSF's from two other cell types: endothelial cells (which line the walls of blood vessels) and fibroblasts (connective-tissue cells that have a role in wound healing).

then it is clear that the CSF's will be of value in other areas as well. For example, bone-marrow transplantation is a difficult procedure that can require weeks of in-hospital care for the recipient and a shorter period for the donor. Administration of CSF's might allow the recipient to recover in as little as a week. Treating donors with CSF's may make it possible to remove much smaller quantities of marrow, allowing donation to be done without hospitalization.

Most of the applications we have discussed entail strengthening a weakened immune system—as in AIDS, cancer chemotherapy or aplastic anemia. More radical applications of the CSF's, however, may come not in such situations but in those where the aim is

to bolster a normal immune system. In the future a multitude of infections may be treated by increasing the number and potency of host-defense cells. Part of the therapy for parasite-caused diseases may be to exploit CSF's in regulating the number of eosinophils. In addition, some experimental cancer therapies involve marking tumor cells with antibodies so that they can be destroyed by monocytes and neutrophils. Such strategies may be enhanced by CSF's that increase the level and activity of these effector cells.

The uniqueness of CSF's in medicine lies in their capacity to make the patient a more formidable defensive entity. In the past the only means available to medicine for improving the host-defense mecha-

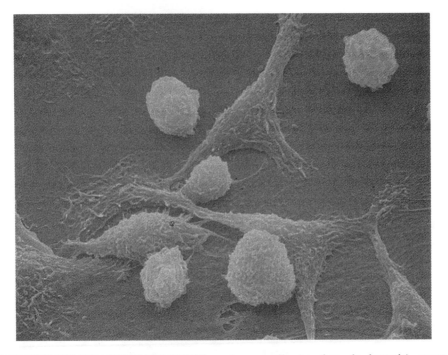

Figure 3.9 MACROPHAGES ATTACK TUMOR CELLS in a micrograph made by S. G. Quan. Both the macrophages (which are flat and irregularly shaped) and the rounded tumor cells come from a mouse. Rather than ingesting tumor cells (as they do bacteria), macrophages kill the tumor cells by releasing toxins such as tumor necrosis factor and by methods (which are still poorly understood) that entail surface contact.

nism were indirect (such as improved nutrition) or were specific to a single disease (such as immunization). A variety of means have been developed to subvert invading organisms, antibiotics being the most dramatic examples. Future research will undoubtedly offer new and more potent ways of interfering with the metabolism of the invader. Thanks to the CSF's, however, clinicians will also have a new strategy based on giving the patient a stronger defense against microbes and even against cancer.

II

THE UNIVERSALS IN INTRACELLULAR CELL COMMUNICATION

. . .

Introduction

There is a remarkable contrast between the diversity of extracellular messengers and their generally restricted targets, and the near universality of a small number of intracellular messengers, and their large number of targets. Chapter 4, "The Molecular Basis of Communication within the Cell," by Michael J. Berridge, presents current global views of the major pathways by which information is conveyed from cell surface to cell interior. It is worth noting that only four pathways are discussed: the classic cyclic adenosine monophosphate (cAMP) messenger system; the phosphatidylinositol-linked ionized calcium (Ca^{++}) messenger system; the receptor tyrosine kinase-linked pathways; and the cyclic GMP messenger system.

What is more remarkable is the fact that the Ca^{++} and cAMP messenger systems do not usually operate as isolated system, but interact in one of several ways to regulate cell function. These interactions are emphasized in Chapter 4, Chapter 5, "Calmodulin," by Wai Yiu Cheung, and Chapter 6, "The Cycling of Calcium as an Intracellular Messenger," by Howard Rasmussen. Thus, for example, in many cases Ca^{++} regulates either cAMP hydrolysis and/or cAMP synthesis, and, conversely, cAMP regulates Ca^{++} influx and/or efflux across the plasma membrane of the same cells, the stimulates the uptake of Ca^{++} by the sarcoplasmic reticulum in cases such as heart muscle cells. Viewed at the level of the intact cells, it is possible to recognize that these two messenger systems can act either in a coordinate, hierarchical, sequential, redundant or antagonistic fashion. Thus, for example, the Ca^{++} and cAMP messenger interact in a synergistic (coordinate) fashion to stimulate insulin secretion, but in an antagonistic fashion in regulating tracheal and vascular smooth muscle contraction. Chapters 4, 5 and 6

represent a current view of the importance, universality, and modes of organization of the Ca^{++} messenger system. This is a field of great research interest, and a number of additional intracellular Ca^{++} binding proteins have been identified, but their functional importance remains to be established.

Chapter 7, "The Molecules of Visual Excitation," by Lubert Stryer, discusses how the cGMP and Ca^{++} messenger systems interact in regulating both smooth muscle contraction and the retinal response to light. What is striking are the operational similarities in the way the retinal cell responds to light and the liver cells to a hormone. The functional coupling of G proteins to hormone or light receptors is a nearly universal attribute of these signalling systems. In Chapters 4 and 7, the effectors to which G proteins couple receptors include cGMP phosphodiesterase, adenylate cyclase, and PI-specific phospholipase C. We now know that G proteins also couple to a variety of other plasma membrane effectors.

Chapter 8, "How Receptors Bring Proteins and Particles into Cells," by Alice Dautry-Varsat and Harvey F. Lodish, focuses on the receptor and emphasizes the fact that it is not a static but a dynamic entity that is continually being synthesized, conveyed to the cell membrane, internalized, and either degraded or returned to the membrane. Of key importance is the fact that internalization is much more efficient when receptor is coupled to the extracellular protein (or ligand). This mechanism operates not only in the case of peptide or protein hormones, but also in the case of a variety of other proteins. One of particular medical importance is the LDL receptor which will be discussed in Chapter 13 in Section III.

The Molecular Basis of Communication within the Cell

*The number of substances serving as signals in cells is remarkably small.
Each such "second messenger" is a crucial guide for the cell,
helping to determine how the cell responds to the organism's needs.*

. . .

Michael J. Berridge
October, 1985

The division of labor among the cells of a multicellular organism requires that each cell population be able to call on the services of some cell populations and respond to the requirements of others. Much of this coordination is achieved by chemical signals. Yet most of the arriving signals never invade the privacy of a cell. Dispersed on the outer surface of the cell are the molecular antennas known as receptors, which detect an incoming messenger and activate a signal pathway that ultimately regulates a cellular process such as secretion, contraction, metabolism or growth. The major barrier to the flow of information is the cell's plasma membrane, where transduction mechanisms translate external signals into internal signals, which are carried by "second messengers."

In molecular terms the process depends on a series of proteins within the cell membrane, each of which transmits information by inducing a conformational change — an alteration in shape and therefore in function — in the protein next in line. At some point the information is assigned to small molecules or even to ions within the cell's cytoplasm. They are the second messengers, whose diffusion enables a signal to propagate rapidly throughout the cell.

The number of second messengers appears to be surprisingly small. To put it another way, the internal signal pathways in cells are remarkably universal. Yet the known messengers are capable of regulating a vast variety of physiological and biochemical processes. The discovery of the identity of particular second-messenger substances is proving, therefore, to be of fundamental importance for understanding how cellular activity is governed. Two major signal pathways are now known. One employs the second-messenger cyclic adenosine monophosphate (cyclic AMP). The other employs a combination of second messengers that includes calcium ions and two substances, inositol triphosphate (IP_3) and diacylglycerol (DG or DAG), whose origin is remarkable: they are cannibalized from the plasma membrane itself (see Figure 4.1).

The paths have much in common. In both of them the initial component, the receptor molecule at the surface of the cell, transmits information

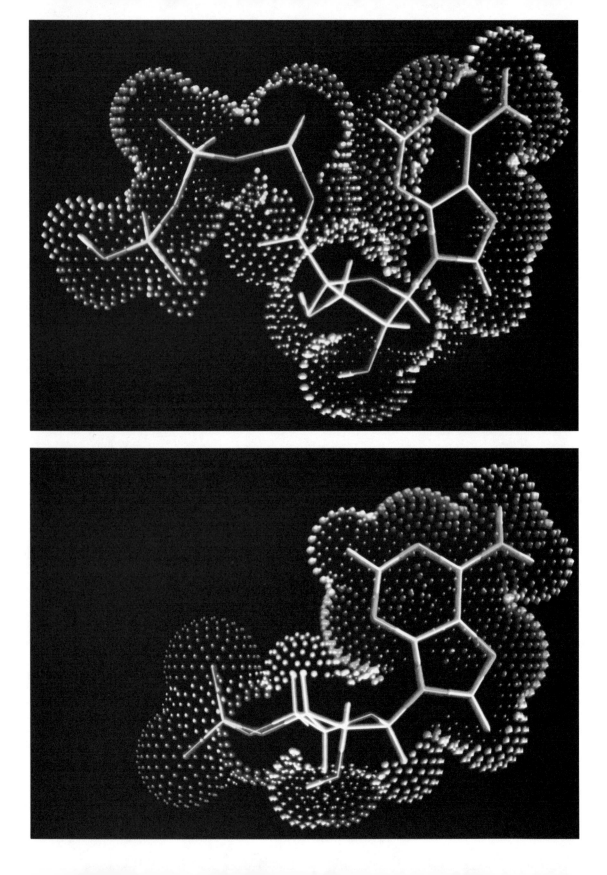

through the plasma membrane and into the cell by means of a family of G proteins: membrane proteins that cannot be active unless they bind guanosine triphosphate (GTP). In both paths the G proteins activate an "amplifier" enzyme on the inner face of the membrane. The enzyme converts precursor molecules into second messengers. As a rule the precursors are highly phosphorylated, that is, rich in phosphate groups (PO_4). For example, the amplifier adenylate cyclase converts adenosine triphosphate (ATP) into cyclic AMP, whereas the amplifier phospholipase C cleaves the membrane lipid phosphatidylinositol 4,5-biphosphate, or PIP_2, into DG and IP_3 (see Figure 4.2).

In both paths, moreover, the final stages are similar: the second messengers induce cellular proteins to change their structure. (In one three-dimensional conformation the protein is inactive; in another it contributes to a cellular function such as secretion or contraction.) There are two main ways in which second messengers function. In one of them the second messenger acts directly. It binds to the protein (specifically, it binds to the "regulatory component" of the protein) and thus triggers a conformational change. A classic example is found in skeletal muscle. There the second messenger calcium binds to the protein troponin C, triggering a conformational change that leads to the contraction of the muscle. In the alternative, more common mechanism the second messenger acts indirectly: it activates an enzyme called a protein kinase, which then phosphorylates a protein. The phosphorylation (that is, the addition of a phosphate group) induces the protein to change its shape.

Of all the steps of the known second-messenger pathways the ones best understood today are the steps of transduction and amplification that ac-

tivate cyclic AMP (see Figure 4.3). The facts emerged in stages, beginning in 1958, when Earl W. Sutherland, Jr., and Theodore W. Rall, working at Case Western Reserve University, discovered cyclic AMP itself. Then in 1971 Martin Rodbell and his colleagues at the National Institutes of Health showed that GTP is essential for the transduction mechanism to generate cyclic AMP. Before information can flow across the membrane two events must occur. At the surface of the cell an external signal must bind to its receptor. Meanwhile, from inside the cell, a GTP molecule must act on its G protein.

The sequence has been elucidated in detail by Alfred G. Gilman and his colleagues at the University of Texas Health Science Center at Dallas. Two types of G proteins turn out to be involved, one of them stimulatory and the other inhibitory. The stimulatory protein, called G_s, links itself to receptors called R_s. The binding of an external signal to such a receptor induces a conformational change in the receptor. The change is transmitted through the cell membrane to G_s, making it susceptible to GTP, which approaches from inside the cell. The binding of GTP to G_s then constitutes an on-reaction: it forces G_s into still another conformation, one that enables it to activate adenylate cyclase and thereby instigate the formation of cyclic AMP. The information carried by the external signal has now been transmitted across the membrane and assigned to an internal signal: a second messenger.

The activity of the G_s-GTP complex is ended by the hydrolysis of the GTP to GDP (guanosine diphosphate); that constitutes the off-reaction. The hydrolysis is catalyzed by the enzyme GTPase. As it happens, the activity of GTPase is inhibited by the toxin produced by the cholera bacillus. The toxin thereby prolongs the life of the G_s-GTP complex, so that the cell produces cyclic AMP continually, even in the absence of an external signal calling for its manufacture. The severe diarrhea characteristic of victims of cholera can be explained in those terms. In the cells of the intestine cyclic AMP is a potent activator of fluid secretion.

The other type of G protein in the cyclic-AMP pathway mediates an inhibitory transduction. The arrival of an external signal at the receptors designated R_i brings on a conformational change in the G protein called G_i (a change again dependent on the binding of GTP); the G protein in turn inhibits adenylate cyclase. Here too the flow of information can be blocked by a bacterial toxin, this one produced by *Bordetella pertussis*, the causative agent of

Figure 4.1 MAKING OF A MESSENGER. The precursor of messenger (*top*) is ATP, which serves the cell by donating energy to chemical reactions and has three parts: adenine forms the upper right of the molecule, its structure dominated by a hexagon and pentagon of carbon atoms (*white*) and nitrogen atoms (*blue*); adenine is joined to ribose, seen almost end on at the bottom of the ATP molecule; ribose is linked to a chain of three phosphate groups, each consisting of a central phosphorus atom (*yellow*) and satellite oxygen atoms (*orange*). For conversion into a messenger ATP is altered into cAMP (*bottom*). Two phosphate groups are removed and the remaining group becomes bound to the rest of the molecule at two of its satellite oxygen atoms.

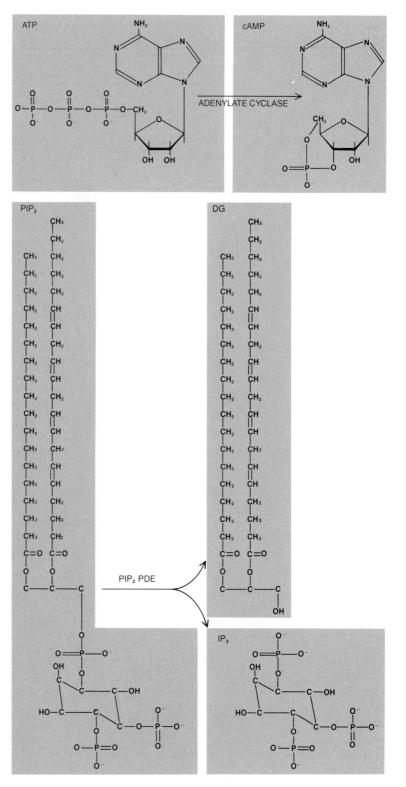

Figure 4.2 CHEMICAL STRUCTURE of identified second messengers is displayed for three messengers: cAMP (see also Figure 4.1), DG and IP₃. Cyclic AMP (*top*) is made in a reaction that cleaves two of the three phosphate groups from ATP. DG and IP₃ (*bottom*) are made from PIP₂ by a simple reaction: the negatively charged "head" of the precursor molecule, a phospholipid found in the inner leaf of plasma membrane, is cleaved from the glycerol backbone that carries the twin fatty acid "tails" of the precursor.

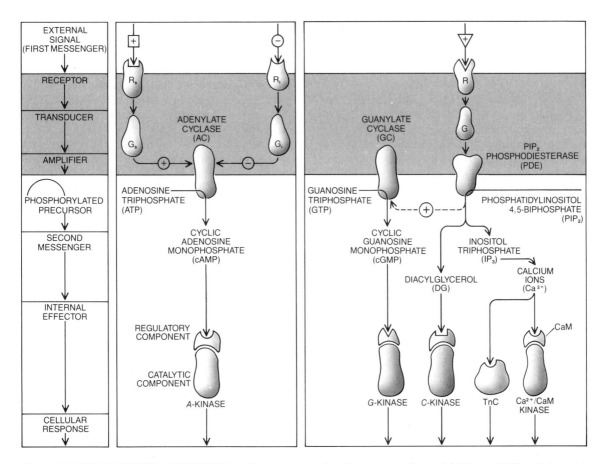

Figure 4.3 KNOWN SIGNAL PATHWAYS in cells share a sequence of events (left). External messengers arriving at receptor molecules in the plasma membrane (gray) activate transducer molecules (G proteins) which in turn stimulate amplifier enzymes that activate internal signals carried by second messengers. The pathway using cAMP (middle) has stimulatory (R$_s$) and inhibitory (R$_i$) receptors, which communicate with adenylate cyclase (AC) by way of G pro-teins. The other pathway (right) uses PDE, which makes PIP$_2$ into a pair of second messengers, DG and IP$_3$. IP$_3$ induces the cell to mobilize Ca^{2+}. The path somehow induces GC to convert GTP to cGMP. In general second messengers bind to the regulatory component of a protein kinase. Calcium binds to proteins including calmodulin (CaM) and troponin C (TnC).

whopping cough. Pertussis toxin blocks the inhibition of adenylate cyclase; however, it is not yet known whether the blockage accounts for any of the symptoms of the disease. Bacterial toxins have been valuable experimental tools for defining the roles of G proteins. Another effective agent is forskolin, an organic molecule isolated from roots of the Indian herb *Coleus forskohlii*. Extracts from the plant are still employed in the folk medicine of India as remedies for ailments including heart diseases, respiratory disorders, insomnia and convulsions. Pharmacological studies have established that forskolin activates adenylate cyclase.

In the cyclic-AMP pathway the final chemical steps are mediated by an A-kinase: a protein kinase that phosphorylates a particular protein when it is activated specifically by cyclic AMP. Each A-kinase has two parts, a catalytic subunit and a regulatory subunit. Cyclic AMP binds to the regulatory subunit, thereby liberating the catalytic one, which is then free to phosphorylate proteins. In fat cells, for example, the enzyme lipase initiates the tapping of the energy content of lipid (fatty) molecules. Hormones such as epinephrine (also called adrenaline) bind to receptors on the cell surface; the receptors, acting through G proteins, influence adenylate cyclase, which makes cyclic AMP; the cyclic AMP stimulates an *A*-kinase, and the *A*-kinase activates lipase by phosphorylating the enzyme. (See Figure 4.4)

Other examples are known in which cyclic AMP works through *A*-kinase to activate cellular enzymes (or cellular processes such as ion transport). As the molecular details of the final steps in the cyclic-AMP pathway are examined in a number of cell types, however, a consistent pattern has begun to emerge: cyclic AMP often proves to serve the cell primarily by activating another second messenger, namely calcium ions. That is, one of the two known signal pathways in cells proves to act chiefly by modulating the other known signal path. The heart provides a now classic example. There epinephrine acts through the cyclic-AMP pathway to modulate the intracellular level of calcium. Thus the force of each heartbeat is governed by a brief calcium pulse. Similar findings (that is, modulation of the calcium pathway by the cyclic-AMP pathway) have emerged in other muscle cells and in a variety of secretory cells, including nerve cells.

The first description of calcium as an intracellular messenger was given in 1883, when the English physician and physiologist Sydney Ringer found that the muscle tissue he was examining failed to contract when the London tap water in his tissue-culture medium was replaced with distilled water. The missing ingredient soon proved to be calcium. A series of observations then showed that calcium regulates not only contraction but also many other cellular processes. It is in fact the predominant second messenger in cells.

Where does the calcium come from? For certain cells, such as neurons, the source is well known: it is the extracellular fluid. Nerve signals arriving at the synaptic terminals of a neuron decrease the voltage difference across the neuronal cell membrane; the resulting "depolarization" opens voltage-sensitive calcium channels through the membrane. Before the depolarization the concentration of free calcium inside the neuron is approximately 1×10^{-7} molar (a value corresponding to some 6×10^{14} calcium ions per centiliter of cytoplasm). The concentration of calcium outside the neuron is four orders of magnitude greater. The depolarization enables calcium ions to flood into the neuron and trigger a cellular response. Even a rather small change in the intracellular concentration of calcium can exert profound changes in cellular activity. In the synaptic termi-

nals of neurons, for example, calcium induces the release of neurotransmitter molecules.

The extracellular fluid cannot, however, be the sole source of calcium ions. For one thing, the absence of extracellular calcium does not prevent the external messenger acetylcholine from stimulating the pancreas to release the digestive enzyme amylase. Thus it has slowly become apparent that the calcium employed by a cell for internal signaling not only enters the cell from outside but also is released from internal reservoirs. There turn out to be many examples of hormones or neurotransmitters employing internal calcium to control physiological processes. The external signal gains access to the internal calcium by stimulating the hydrolysis of a lipid molecule that is part of the plasma membrane.

In 1953 Mabel N. and Lowell E. Hokin, working at the Montreal General Hospital, observed that the administration of acetylcholine to secretory cells of the pancreas increased the incorporation of radioactive phosphate groups (PO_4 groups containing phosphorus 32) into phosphatidylinositol (PI), one of the phospholipid constituents of cell membranes. Like other membrane lipids it has a hydrophobic part (two fatty acid chains attached to glycerol) linked to a hydrophilic part, in this case the "head group" inositol phosphate. Stimuli such as acetylcholine cause it to be cleaved into these two components. The increased incorporation of phosphorus 32 observed by the Hokins was a secondary event due to the subsequent resynthesis of PI.

The key point was that an external signal had been found to stimulate the turnover (the hydrolysis and resynthesis) of a membrane lipid. The Hokins proposed that the increased turnover had something to do with the mechanism of exocytosis by which the cells of the pancreas release digestive enzymes. Subsequent studies established a broader conclusion: the increased turnover occurred in response to many stimuli, not necessarily the ones that activate secretion. Hence the impression grew that the turnover of membrane lipids plays a more general role in the life of cells.

In 1975 Robert Michell of the University of Birmingham suggested such a role. Noting a strong correlation between the ability of an external signal to stimulate inositol-lipid turnover and the mobilization of calcium inside the cell, he suggested that the change in lipid turnover triggered by external signals is responsible for generating internal calcium signals. Michell and his colleagues made a further suggestion. Cell membranes contain three inositol lipids, but only one of them, the relatively minor inositol lipid phosphatidylinositol-4,5-biphosphate, or PIP$_2$, seemed to change (in particular, it was hydrolyzed) as part of the mechanism mobilizing calcium.

How might the hydrolysis of a particular membrane lipid act to increase the intracellular concentration of calcium? It has taken almost a decade to work out the details. In order to understand how second messengers can emerge from cell membrane and how they might function, a digression is required: I must turn to some basic aspects of the biochemistry of the molecules composing cell membranes. Placed in an aqueous medium, phospholipids, of which the inositol lipids are notable examples, spontaneously coalesce into the orderly double-layered alignment that constitutes the basic structure of a biological membrane. The traditional view is that the lipid bilayer functions as an inert, permeable barrier. Hence phospholipids have often been dismissed as playing a rather passive role in the life of the cell. A role for a membrane phospholipid in an intracellular signal pathway comes, therefore, as something of a surprise.

Phosphatidylinositol is a typical phospholipid, situated primarily in the inner leaflet of the bilayer. The remarkable thing about it is that it gets converted into PIP$_2$, an unusual phospholipid that has three phosphate groups instead of the one group found in all other membrane lipids. The additional phosphates, derived from ATP, are added sequentially and specifically to the carbon-4 and -5 positions of the six-carbon ring in inositol.

PIP$_2$ is the inositol lipid that interested Michell. In response to external signals it is hydrolyzed into diacyglycerol and inositol triphosphate, or IP$_3$ (see Figure 4.5). Two groups of investigators, Richard Haslam and Monica Davidson of McMaster University in Ontario and Shamshad Cockcroft and Bastion D. Gomperts of University College London have found that GTP is a necessary part of the sequence. Again, therefore, it seems that a G protein couples surface receptors to an amplifier (in this case the enzyme phospholipase C). Ultimately the diacyglycerol and IP$_3$ are recycled, the first by a series of reactions composing what is called the lipid cycle, the second by a series of reactions called the inositol phosphate cycle. The two cycles merge, reconstituting phosphatidylinositol.

The final step in the inositol phosphate cycle is particularly interesting; it is the conversion of inositol monophosphate (IP$_1$) into free inositol by the enzyme inositol 1-monophosphatase. Working in St. Louis at the Washington University School of Medicine, James Allison and William R. Sherman have shown that the action of the enzyme is inhibited by lithium ions. The administration of lithium may therefore slow the rate of resynthesis of phosphatidylinositol, impairing the effectiveness of any neuronal mechanisms that depend on the inositol lipids to carry signals. Perhaps that accounts for the efficacy of lithium ions in controlling manic-depressive mental illness.

It was while measuring the rate at which inositol phosphates form in the salivary gland of the fly in response to the hormone serotonin that my attention was first drawn to IP$_3$. Analysis of the water-soluble metabolites extracted from the gland revealed the presence of at least four distinct substances. One was free inositol; the others turned out to be the inositol phosphates IP$_1$, IP$_2$ and IP$_3$. The analysis was done in collaboration with Robin Irvine and Rex Dawson of the Agricultural Research Council's Institute of Animal Physiology in Babraham and Peter Downes of the Medical Research Council's Neurochemical Pharmacology Unit in Cambridge. In 1964 Dawson, who had been studying the enzyme that hydrolyzes PIP$_2$, stored some of the reaction product, IP$_3$, in a deep freeze at Babraham. Almost two decades later it served as a standard by which to verify our identification of the fly-gland metabolite.

By comparing the rate of IP$_3$ formation with the rate of onset of physiological signs that accompany the secretion of fluid by cells of the insect salivary gland, John P. Heslop and I were able to show that the administration of serotonin brings on the generation of IP$_3$ at a rate fast enough so that IP$_3$ could conceivably function as a second messenger: it would mobilize calcium, which in turn would cause the secretion of saliva. Seeking direct evidence of such a function, we sent a sample of IP$_3$, which Irvine had prepared from red blood cells, to the Max Planck Institute in Frankfurt. There Hanspeter Streb and Irene Schulz found that when the sample was applied to cells from the pancreas of a rat, it caused a profound release of calcium. (The cells had first been permeabilized so that the IP$_3$ applied by the investigators could gain access to the interior of the cells.)

This first demonstration of a release of calcium induced by IP$_3$ has now been confirmed in a num-

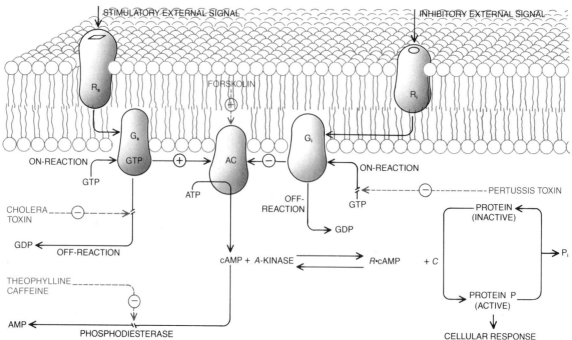

STIMULATORY EXTERNAL SIGNALS	INHIBITORY EXTERNAL SIGNALS	TISSUE	CELLULAR RESPONSE
ADRENALINE (BETA RECEPTORS)		SKELETAL MUSCLE	BREAKDOWN OF GLYCOGEN*
ADRENALINE (BETA RECEPTORS)		FAT CELLS	INCREASED BREAKDOWN OF LIPIDS
ADRENALINE (BETA RECEPTORS)		HEART	INCREASED HEART RATE AND FORCE OF CONTRACTION*
ADRENALINE (BETA RECEPTORS)		INTESTINE	FLUID SECRETION*
ADRENALINE (BETA RECEPTORS)		SMOOTH MUSCLE	RELAXATION*
THYROID STIMULATING HORMONE		THYROID GLAND	THYROXINE SECRETION
VASOPRESSIN (V_2 RECEPTORS)		KIDNEY	REABSORPTION OF WATER
GLUCAGON		LIVER	BREAKDOWN OF GLYCOGEN*
SEROTONIN		SALIVARY GLAND (BLOWFLY)	FLUID SECRETION
PROSTAGLANDIN I_1		BLOOD PLATELETS	INHIBITION OF AGGREGATION AND SECRETION *
	ADRENALINE ($ALPHA_2$ RECEPTORS)	BLOOD PLATELETS	STIMULATION OF AGGREGATION AND SECRETION *
	ADRENALINE ($ALPHA_2$ RECEPTORS)	FAT CELLS	DECREASED LIPID BREAKDOWN
	ADENOSINE	FAT CELLS	DECREASED LIPID BREAKDOWN

Figure 4.4 DETAILS OF SIGNAL PATHWAYS. In the cAMP pathway (*left*) signals from stimulatory (R_s) and inhibitory (R_i) receptors converge on the amplifier enzyme AC, which converts ATP into cAMP. *G* proteins are activated by GTP (*on-reaction*) and curtailed when the GTP is hydrolized (*off-reaction*) to GDP. For its part, cAMP binds to the regulatory component (R) of its protein kinase, liberating the catalytic component (C), which is then free to phosphorylate specific proteins that regulate a cellular response. Drugs affecting a particular stage in the sequence are shown in color. Some known cellular responses are listed. In many cases (*asterisks*) cAMP proves to modulate the activity of calcium, which in turn governs the response. In the inositol-lipid path (*right*) external signals

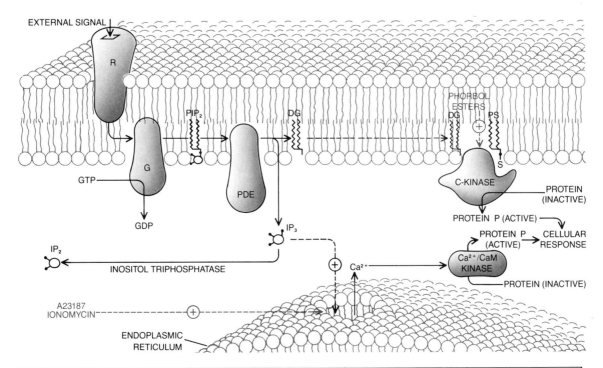

EXTERNAL SIGNAL	TISSUE	CELLULAR RESPONSE
VASOPRESSIN	LIVER	BREAKDOWN OF GLYCOGEN
ACETYLCHOLINE	PANCREAS	AMYLASE SECRETION
ACETYLCHOLINE	SMOOTH MUSCLE	CONTRACTION
ACETYLCHOLINE	OOCYTES (XENOPUS)	CHLORIDE PERMEABILITY
ACETYLCHOLINE	PANCREATIC BETA CELLS	INSULIN SECRETION
SEROTONIN	SALIVARY GLAND (BLOWFLY)	FLUID SECRETION
THROMBIN	BLOOD PLATELETS	PLATELET AGGREGATION
ANTIGEN	LYMPHOCYTES	DNA SYNTHESIS
ANTIGEN	MAST CELLS	HISTAMINE SECRETION
GROWTH FACTORS	FIBROBLASTS	DNA SYNTHESIS
LIGHT	PHOTORECEPTORS (LIMULUS)	PHOTOTRANSDUCTION
SPERMATOZOA	SEA URCHIN EGGS	FERTILIZATION
THYROTROPIN RELEASING HORMONE	ANTERIOR LOBE OF PITUITORY GLAND	PROLACTIN SECRETION

bind to receptors (R), which transmit information through a *G* protein (G) to activate PIP_2 phosphodiesterase (PDE). In turn PDE cleaves PIP_2 into IP_3 and DG. The IP_3 is water-soluble, and so it diffuses into the cytoplasm. There it releases calcium from storage in the membranous intracellular caverns called endoplasmic reticulum. In turn the calcium stimulates a protein kinase. The DG remains in the membrane, where it activates *C*-kinase; the membrane phospholipid called phosphatidyl serine (PS) is a cofactor for the activation. The two limbs of the pathway lead to the phosphorylation of distinct sets of proteins. The limbs can be activated independently by means of the drugs indicated in color.

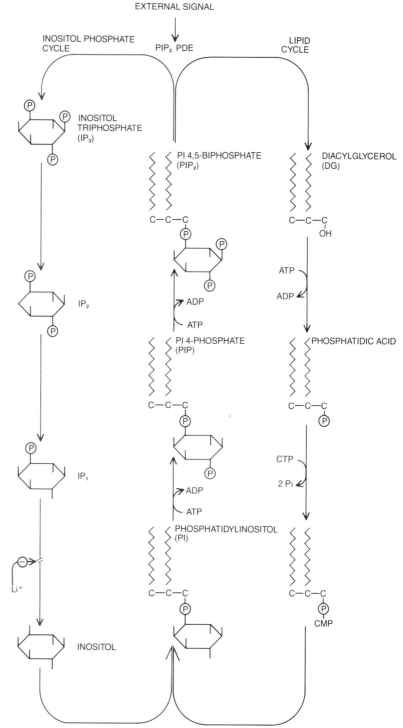

Figure 4.5 INOSITOL-LIPID CYCLES replenish the supply of second messengers made from inositol lipids. External signals act through the enzyme PIP_2 phosphodiesterase, which cleaves PIP_2 into the messengers DG and IP_3. The two are then directed through a sequence of chemical reactions that prepare them to be rejoined, forming phosphatidylinositol (PI) and ultimately remaking the PIP_2. The cycles require the continuous presence of ATP and cytosine triphosphate (CTP), which are sources of phosphate groups (P_i). Among the few drugs known to affect a part of the path, lithium (*color*) is notable: it blocks the conversion of IP_1 into the free inositol required for the synthesis of PI.

ber of different types of cells. Gillian Burgess and James W. Putney, Jr., of the Medical College of Virginia and John R. Williamson of the University of Pennsylvania School of Medicine have shown, for example, that IP$_3$ mobilizes stored calcium as part of the hormonal mechanism for releasing glucose from the liver. Moreover, Yoram Oron and his colleagues at Tel-Aviv University have shown that the current of chloride ions induced by acetylcholine in immature oocytes, or egg cells, of the frog *Xenopus* can be duplicated by injecting the cells with IP$_3$. In addition IP$_3$ elicits many of the early events of fertilization. For example, granules stored just under the surface of the egg are normally secreted within minutes of fertilization to form the thick protective layer called the fertilization membrane. The secretion, which is calcium-dependent, can be triggered by injecting eggs with IP$_3$. In each case the IP$_3$ proves to act predominantly by causing the release of calcium from imprisonment in the cell's endoplasmic reticulum, an internal membrane that forms a system of caverns inside the cell. In turn the calcium elicits the cellular response.

The discovery of the second messenger IP$_3$ has led to speculation that IP$_3$ may function as a second messenger in skeletal muscle. In muscle the depolarization of the infoldings of muscle membrane known as transverse tubules somehow triggers the release of calcium from the sarcoplasmic reticulum, a structure analogous to the endoplasmic reticulum of nonmuscle cells. The calcium triggers muscle contraction. Roger Y. Tsien and Julio Vergara of the University of California at Berkeley and Tullio Pozzan and his colleagues at the University of Padua have found that isolated muscle fibers contract in response to the administration of IP$_3$. The idea, then, is that in muscle IP$_3$ functions as the link between depolarization and calcium. The verification of the speculation would be the jewel in the crown of the IP$_3$ hypothesis; the problem of how calcium signals are generated in skeletal muscle has puzzled physiologists for decades.

A further reason the inositol-lipid transduction mechanism is attracting much interest is that the signal pathway bifurcates. One product of the hydrolysis of the inositol lipid PIP$_2$ is IP$_3$, whose role I traced above. The other product, diacylglycerol, remains in the membrane, yet apparently it functions, like IP$_3$, as a second messenger. Yasutomi Nishizuka and his colleagues at Kyoto University propose that it activates a membrane-bound protein kinase, which they have named C-kinase (see Figure 4.4).

The contribution of each limb of the bifurcating inositol-lipid path can be assessed with pharmacological agents that mimic the action of a particular second messenger and therefore stimulate only one limb of the path. The phorbol esters (substances found in the oil expressed from the seed of the small Southeast Asian tree *Croton tiglium*) mimic the action of diacylglycerol by acting directly on C-kinase. (The phorbol esters cause inflammation of the skin and are potent tumor-inducing agents when they are applied to experimental animals in combination with a carcinogen.) On the other hand, calcium ionophores (molecules that shield the electric charge of a calcium ion and smuggle it across the cell membrane) mimic the action of IP$_3$ by introducing free calcium into the cell. The studies establish that the two limbs are synergistic: in blood platelets, for example, Nishizuka has found that the combination of a phorbol ester and a calcium ionophore induces a maximal secretion of serotonin at doses that have no effect when each drug is administered alone.

The importance of the overall two-branched signal pathway is hard to overstate: a great many cellular processes can be switched on experimentally by the combined administration of a phorbol ester and a calcium ionophore. Perhaps the most notable finding is that the synthesis of DNA can be initiated: the finding is notable because it suggests that the signal pathways responsible for routine cellular activities such as secretion and contraction may also regulate growth. The action of phorbol esters as tumor-promoting agents is probably based, for example, on their ability to amplify the DG/C-kinase limb of the inositol-lipid signal pathway. Indeed, the prospect arises that alterations of intracellular signal pathways may be a cause of cancer, in which the normal regulation of cell growth is disrupted.

Cells grow by progressing through the stages of the cell cycle. In an initial phase they increase in size. This is the first growth phase, G$_1$. Next they replicate all their chromosomes (during S, the DNA-synthesis phase) and prepare themselves for cell division (during G$_2$, the second growth phase). Finally they divide (during M, the mitosis phase). Just after cell division comes a branch point. Each daughter cell arising from the division can reenter the cell cycle and hence can divide again. Alternatively, a daughter cell can enter the G$_0$ phase, during which it differentiates, becoming capable of performing some specialized task in one of the tissues of the body. For certain types of cells, such as neurons, differentiation puts an end to division; for other types the cell has the option of returning to the cell cycle to engender further progeny. In the latter case the return to the cell cycle is determined by the action of growth factors: substances released by one group of cells that stimulate the growth of others.

Just how growth factors act is still very much a mystery. Clearly, however, they must instigate the sending of signals from the cell surface (where the factors act by binding to receptors) to the nucleus (where DNA is replicated). I shall consider two possible pathways (see Figure 4.6). One of them is not yet well understood. Employed by growth factors such as insulin and epidermal growth factor (EGF), it appears to rely on receptors that activate the enzyme tyrosine kinase. The pathway may be in essence a cascade based on the phosphorylation of a succession of proteins.

The other signal pathway, employed by such factors as platelet-derived growth factor (PDGF), appears to be identical with the pathway employed by hormones and neurotransmitters. The PDGF arriving at the surface of a cell stimulates the hydrolysis of PIP_2 into the second messengers IP^3 and diacylglycerol, which may then contribute to the events that make up growth phase G_1 and prepare the cell for DNA synthesis. More specifically, IP_3 seems to act by mobilizing intracellular calcium; diacylglycerol activates C-kinase, which in turn activates a membrane-bound ion-exchange mechanism. The mechanism extrudes protons (hydrogen ions) from the cell, thus raising intracellular pH. Together the activation of calcium and the raising of pH are thought to contribute to the synthesis of RNA and protein that prepares the cell to synthesize DNA.

Since each signal pathway consists of a sequence of reactions controlled by specific proteins (receptors and enzymes), the genetic material of the cell must include the genes responsible for the synthesis of the proteins the pathways require. Any aberration of the function of such genes might lead, therefore, to abnormalities of cellular growth, and conceivably to the uncontrolled growth and structural transformations typical of cancer. About 25 remarkable genes have in fact been identified: genes whose inappropriate function has been linked to the incidence of cancer. They are collectively termed oncogenes. Until recently the precise normal function of each such gene was obscure. It is now apparent that

some of them encode the structure of various components of the signal pathways controlling cell growth.

The first link between an oncogene and a component of an intracellular signal pathway was made simultaneously by two groups, one headed by Russell F. Doolittle of the University of California at San Diego, the other by Michael Waterfield of the Imperial Cancer Research Fund Laboratories in London. The groups discovered that the oncogene called *sis* controls the synthesis of platelet-derived growth factor. Discoveries of similar import followed at other laboratories. The *erb b* gene proved to encode the structure of a protein almost identical with the epidermal-growth-factor receptor. The receptor has three main parts. An external part, exposed at the surface of the cell, includes the EGF-binding domain; a middle part spans the cell membrane; an inner part, exposed to the cytoplasm, expresses the protein-phosphorylating activity of a tyrosine kinase. The product of the *erb b* gene is a truncated version of the receptor, a version that lacks the external part of the protein. Conceivably the truncated version initiates signals inside the cell even in the absence of EGF.

The *ras* oncogene also fits the pattern. It is known to be active in many types of cancer cell. Its function is not yet known, but its product has characteristics of a G protein: it is a constituent of the cell membrane and it binds and hydrolyzes GTP. One possibility is that it may intervene between growth-factor receptors and such signal amplifiers as the phosphodiesterase enzyme that cleaves PIP_2 into the second messengers IP_3 and diacylglycerol. Two oncogenes, *src* and *ros*, seem to have a role in the conversion of phosphatidylinositol into PIP_2. That is, they appear to regulate the enzymes that replenish the precursor of the inositol-lipid second messengers. Two other oncogenes, *myc* and *fos*, apparently function at the other end of the intracellular signal cascade. Philip Leder and his colleagues at the Harvard Medical School have found that the abundance of messenger RNA transcribed from the *myc* oncogene increased greatly within an hour of treating fibroblasts (immature connective, tissue cells) with PDGF. Transcripts of the *fos* oncogene appear even earlier. The *myc* and *fos* genes specify the structure of proteins found in the nucleus and so may prove to take part in the sequence of events initiating the synthesis of DNA.

One begins to see that an integrated network of

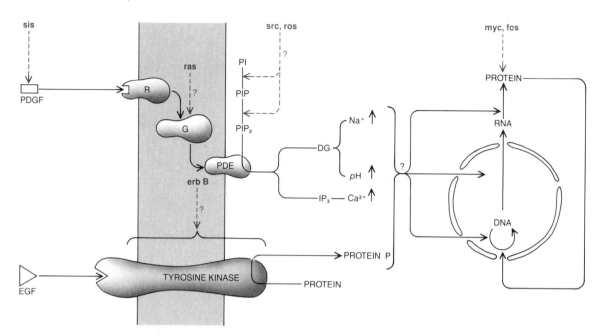

Figure 4.6 REGULATION OF CELL GROWTH is presumably a function of second-messenger pathways. The central problem is to explain how growth factors, which act on cell-surface receptors (*left*), can instruct the machinery in the nucleus (*right*) to begin DNA synthesis. Some growth factors, including PDGF, appear to employ the inositol-lipid pathway: IP_3 mobilizes calcium, whereas DG activates a membrane-bound pump that exchanges protons for sodium ions, thus increasing intracellular *p*H and the concentration of sodium ions. Each change could convey information to the nucleus. Other growth factors, including EGF, appear to employ a different pathway. The EGF receptor includes an inner part that phosphorylates proteins. A number of oncogenes may act (*color*) by disrupting the growth-control pathways. The *erb b* oncogene specifies a protein identical with part of the EGF receptor. Question marks indicate speculative possibilities.

oncogene products is responsible for conveying information from the cell surface to the nucleus. Some oncogenes (*sis*) specify growth factors, which act by inducing other oncogenes (*myc* and *fos*) to produce substances active within the nucleus. Distortions of such sequences lead to uncontrolled cell growth and cancer.

In all the foregoing I have neglected a further second-messenger candidate. It is cyclic GMP, which differs structurally from cyclic AMP in that guanosine takes the place of adenosine. Although cyclic GMP has the hallmarks of a second messenger, its precise role in the cell is not well understood. In the first place, guanylate cyclase, the enzyme that makes cyclic GMP from GTP, is usually not connected to a receptor. Nevertheless, the formation of cyclic GMP often occurs together with the activation of the inositol-lipid pathway. Apparently some molecule created by the hydrolysis of inositol lipids brings on the formation of cyclic GMP. The end of the signal pathway is equally obscure. Cyclic GMP is known to activate a protein kinase (in particular the one called G-kinase), which in turn phosphorylates certain proteins. Their functions are not known.

Still, cyclic GMP has some striking effects, which have been demonstrated best in the nervous system. For example, James W. Truman of the University of Washington has uncovered a role for cyclic GMP in controlling a complex pattern of insect behavior. At the end of metamorphosis moths escape from their cocoons by means of a carefully orchestrated pattern of writhing and wriggling triggered by an eclosion (hatching) hormone released from the brain. This preprogrammed behavioral pattern is initiated by an increase in the level of cyclic GMP, which occurs when eclosion hormone acts on the nervous system.

Another tissue in which a clear function for cyclic

GMP is beginning to emerge is the retina. Specifically, the function is emerging in the vertebrate photoreceptors, or light-sensitive cells, known as rod cells. A rod cell is a sensory transducer stationed between the visual world and the brain. It is an elongated cell. At one end it receives photons, or quanta of light; at the other end it releases a neurotransmitter, thus dispatching signals to neurons. Two things about the sequence are remarkable. First, the amount of transmitter released by the cell is greatest in the absence of light. This suggests a curious attribute of rod cells: in response to external signals (in this case photons) the internal messenger must decrease its activity. Second, the receptor in rod cells is the molecule rhodopsin, which is present in an elaborate stack of disks of membrane inside the cell. On the other hand, the release of neurotransmitter is regulated by changes in the permeability of the plasma membrane to sodium ions. That defines the central problem in phototransduction: What is the identity of the second messenger that carries information from the internal disks to the plasma membrane?

For some time two camps of investigators have held opposing views about the identity of the messenger. One camp championed calcium; the other championed cyclic GMP. The truth may lie in the middle: both may take part. Here I shall concentrate on cyclic GMP. The current hypothesis is that sodium channels through the plasma membrane of the rod cell are kept open in the dark by a high intracellular level of cyclic GMP. A surprising aspect of the hypothesis has been reported by Evgenii Fesenko and his colleagues at the Institute of Biological Physics in Moscow. Cyclic GMP appears to open the channel directly, without activating a protein kinase. When photons arrive, they are absorbed by rhodopsin, which in response induces a molecule called transducin (another member of the G-protein family) to bind GTP and activate the enzyme called cyclic-GMP phosphodiesterase. The result is a precipitous fall in the level of cyclic GMP and the closure of sodium channels. It should be noted, however, that in the photoreceptors of the crab *Limulus* the effect of light can be duplicated by the injection of IP$_3$ but not of cyclic GMP. Perhaps the second messenger in visual signal transduction varies from species to species.

Interest in cyclic GMP is likely to grow now that Ferid Murad of the Stanford University School of Medicine has shown that atrial natriuretic factor, a newly discovered hormone secreted by the atrium of the heart, seems to relax the smooth muscle surrounding blood vessels (and so take part in regulating blood pressure) by increasing the level of cyclic GMP. Like cyclic AMP, cyclic GMP may act by modulating the action of calcium.

Have all the second-messenger signal pathways been identified? The answer is almost certainly no. There are external signals that induce profound effects in cells by way of a signal pathway that remains totally obscure. An intriguing example is offered by the maturation of starfish oocytes. The administration of 1-methyl adenine (the maturation hormone released from the surrounding follicle cells) to the surface of starfish oocytes causes dissolution of the nucleus and the reinitiation of meiosis: the process of cell division by which sex cells such as oocytes increase in number. Just how a simple substance such as 1-methyl adenine acting at the cell surface can make the nucleus disappear is a mystery. Another example is offered by insulin. The pathway by means of which it drives lipid and glycogen synthesis in muscle and liver cells is a mystery. Joseph Larner and his colleagues at the University of Virginia School of Medicine have proposed that insulin may act through a peptide (a short amino acid chain), but the evidence is far from complete. All that seems certain is that the insulin receptor, like the receptor for epidermal growth factor, acts as a tyrosine kinase. Although some of the signal pathways in cells have now been mapped out, it seems clear that uncharted pathways remain.

Calmodulin

*Calcium ions are important intracellular messengers. Often their
message is relayed by calmodulin, a ubiquitous protein that binds
calcium ions and is thereby activated to regulate the function of various enzymes.*

. . .

Wai Yiu Cheung
June, 1982

For an organism to function its cells must communicate with one another. They do so through direct contact or through signals delivered by an electrical impulse or a chemical messenger. The cells are divided into various compartments and subcellular organelles, and these too must communicate. Any signal requires a receiver; for a chemical messenger the receiver is typically a protein that senses the arrival of the messenger and interprets the message by regulating the appropriate cellular activity. Perhaps the most versatile intracellular messenger is the calcium ion. Its major receptor, which appears to mediate and modulate most of the ion's manifold activities, is the ubiquitous protein called calmodulin.

The importance of the calcium ion as a cellular regulator began to be appreciated as long ago as 1883, when the British physiologist Sydney Ringer learned that muscle contraction can be maintained in an isolated frog heart only if the ion is present in the medium bathing the heart. He went on to show that many other physiological activities also require calcium. In the 1950's L. V. Heilbrunn of the University of Pennsylvania showed that injecting the ion into a muscle fiber causes the fiber to contract. Since then it has become clear that the calcium ion

affects almost all aspects of cellular physiology. (It is the ionic form of calcium, Ca^{++} or Ca^{2+}, that is active under physiological conditions because in the watery internal environment the calcium atom gives up two electrons and becomes a divalent cation: an ion with two units of positive electric charge.)

I have mentioned calcium's role in muscle contraction. It also mediates endocytosis and exocytosis (the intake and output of substances through the cell membrane), the motility of cells, the movement of chromosomes prior to cell division and perhaps the division process itself. It has a central role in the metabolism of glycogen, the storage form of glucose. And it has an influence on both the synthesis and the release of neurotransmitters, the molecules that carry a signal from one nerve cell to another. Notwithstanding these many prominent functions of calcium, however, its mechanism of action remained obscure until quite recently.

The first suggestion of how the ion works at the molecular level came in the 1950's from investigations of muscle activity. It was found that in skeletal and cardiac muscle (which is called striated muscle because of its striped appearance in a micrograph) calcium binds to a protein called troponin C. It was still not known how calcium causes contraction in smooth muscle (most involuntary muscle) and contractile movements in nonmuscle cells, or how it works in other tissues. The answers to many of these questions have involved calmodulin.

In the late 1950's Earl W. Sutherland, Jr., of the Western Reserve University School of Medicine discovered the mechanism whereby the signal delivered by a hormone to a cell is received and turns on a cellular function. The hormone was glucagon, the cells were those of the liver and the function was the breakdown of stored glycogen to supply glucose when the body is under stress. Sutherland found that the binding of glucagon to a receptor on the cell's outer membrane activates an enzyme, adenylate cyclase, embedded in the membrane and extending through it. The adenylate cyclase converts adenosine triphosphate (ATP) in the cytoplasm of the cell into the specialized nucleotide cyclic adenosine monophosphate (cyclic AMP). The cyclic AMP serves as an intracellular messenger that relays the hormone's message to the biochemical machinery of the cell, causing it to break down glycogen. Soon cyclic AMP was shown to function as an intracellular messenger in a wide variety of cells that respond to various hormones.

In 1964, as a postdoctoral fellow at the Johnson Research Foundation of the University of Pennsylvania, I was studying variations in the intracellular concentration of cyclic AMP. I needed to understand the regulatory properties of the enzyme phosphodiesterase, which breaks down cyclic AMP to the nucleotide 5'-adenosine monophosphate and thus terminates its signal. In order to purify a phosphodiesterase from the bovine brain (a rich source of the enzyme) I passed a crude brain extract through an ion-exchange chromatography column (see Figure 5.1). The negatively charged resin in such a column binds different components of a crude extract more or less strongly, so that when an appropriate solution is subsequently poured through the column and successive fractions are collected, the component being sought is concentrated in certain fractions. If the component is an enzyme such as phosphodiesterase, one expects those fractions to show more enzyme activity than the others.

To my surprise the partially purified phosphodiesterase showed substantially less enzyme activity than crude brain extract. I repeated the ion-exchange procedure and the assays for enzyme activity and could find no error. It was a peculiarity of the assay procedure that led to the solution of the puzzle and to the discovery of calmodulin. I was doing the assay in two stages, first treating cyclic AMP with phosphodiesterase and then treating the 5'-adenosine monophosphate product with snake venom, which includes an enzyme that yields adenosine and phosphate. The quantity of phosphate was then measured to assess phosphodiesterase activity.

On the basis of past experience the assay results should have been the same whether the two-stage procedure was followed or the assay was done in one step, by mixing the snake venom with the phosphodiesterase. Because I was baffled by the loss of phosphodiesterase activity I repeated the assay carefully, doing it in one stage as well as in two stages. In the case of the crude brain extract there was no difference: the same level of enzyme activity was indicated by both procedures. In the case of the purified enzyme, however, there was a difference: the one-stage assay revealed significantly more activity. The two-stage assay had shown a loss of phosphodiesterase activity because snake venom was not present during the phosphodiesterase reaction. When I added snake venom to the purified enzyme, the phosphodiesterase activity increased; venom added to the crude extract had no effect.

One possible explanation of these results was that an activating factor in the crude extract was removed in the purification procedure and the snake venom had some effect that made up for the factor's absence. In 1967 (by which time I had moved to St. Jude Children's Research Hospital in Memphis) I was able to report that one of the chromatographic fractions did indeed include a factor required for phosphodiesterase activity. The factor was a protein, but it was clearly not phosphodiesterase itself. Adding the protein to the purified, inactive enzyme restored enzyme activity; adding it to the crude brain extract, where it presumably was already present, did not increase activity (see Figure 5.2). The fact that bovine brain phosphodiesterase is fully active only in the presence of a protein activator that can be removed from the enzyme was soon confirmed by other workers, notably Shiro Kakiuchi of Osaka University.

The protein was calmodulin, but it got that name only much later, after its relation to calcium was well established. In the early days it was clear only that the protein has no enzymatic activity of its own, that it is remarkably stable when subjected to

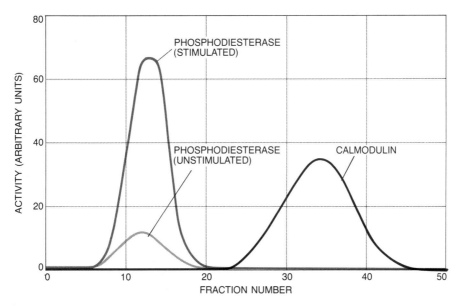

Figure 5.1 PURIFICATION of the enzyme phosphodiesterase from an extract of bovine brain tissue separates the enzyme from calmodulin and thereby reduces the enzyme's activity. The extract is passed through an ion-exchange column. Proteins adsorbed onto the resin in the column are eluted (washed out) more or less quickly depending on their electric charge. The eluted proteins are collected in successive fractions whose activity is measured. The curves show phosphodiesterase activity (*color*) and calmodulin activity (*black*). The purified enzyme, unstimulated by calmodulin, has low activity. When either calmodulin or snake venom (a reagent in the assay for phosphodiesterase) is added to phosphodiesterase fractions, it stimulates enzyme.

heat and that it interacts with phosphodiesterase stoichiometrically (in specific proportions) rather than catalytically (like an enzyme). In our laboratory and others methods were developed for purifying the protein and for characterizing it biochemically. In 1973 Jerry H. Wang of the Faculty of Medicine of the University of Manitoba found that calmodulin binds calcium. This explained some earlier observations. Kakiuchi had noted that the brain phosphodiesterase requires calcium ions for full activity and that calmodulin increases the enzyme's response to the presence of calcium; I had found a small amount of calcium to be associated with a partially purified phosphodiesterase that retained some calmodulin. Soon the relation of calmodulin and calcium was clarified. Calmodulin mediates calcium's effect on phosphodiesterase; the function of the ion is to activate calmodulin, which in turn activates phosphodiesterase.

With calmodulin under investigation in a number of laboratories it was not long before the protein was characterized in detail. It is a single chain of 148 amino acid subunits, with a molecular weight of 16,700 (see Figure 5.3). A third of the subunits are either glutamate or aspartate, amino acids with acidic side chains that supply negatively charged carboxylate (COO^-) groups; it is primarily these groups that bind to the calcium ion. The protein appears to be folded into four roughly matching domains, each of which has a calcium-binding site. The affinity of the sites for calcium is such that whether the ion is bound or not depends on its concentration in the intracellular environment. That degree of affinity is an essential property for a signal-sensing receptor.

Calmodulin is a notably tough molecule. It withstands a very low (acidic) pH and boiling water, conditions that disable most proteins. Within cells calmodulin is found free in the cytoplasm or associated with membranes and organelles. The steady-state concentration of the protein varies with the kind of cell and with conditions. The important thing is that the concentration does not seem to limit the rate of enzymatic reactions, that is, there always seems to be more than enough calmodulin.

In addition to the high proportion of glutamate and aspartate, calmodulin is notable for the com-

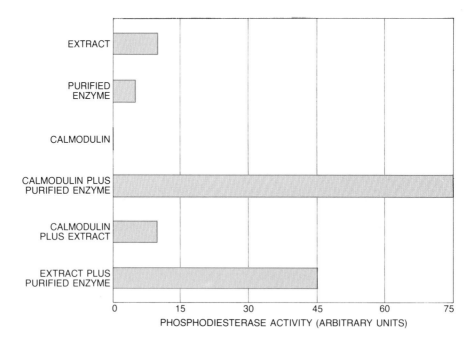

Figure 5.2 PHOSPHODIESTERASE ACTIVITY is reconstituted with calmodulin. Phosphodiesterase purified in the chromatographic column has less activity than the crude brain extract. The calmodulin fraction has no phosphodiesterase activity. Adding calmodulin to the purified enzyme markedly increases enzyme activity. Adding calmodulin to the brain extract has no effect because the extract includes enough calmodulin to stimulate the enzyme. The presence of ample calmodulin in the extract is confirmed by mixing extract with purified phosphodiesterase. Mixture shows a level of activity much higher than sum of activities measured separately.

plete lack of two easily oxidized (and therefore easily degraded) amino acids, tryptophan and cysteine. The lack of cysteine (which includes a sulfur atom) means that disulfide bridges cannot be formed between various regions of the chain. The absence of such bridges, together with the lack of hydroxyproline (a modified amino acid that often braces a turn of the protein chain), makes calmodulin very flexible. An ability to resume its normal configuration after being distorted by harsh conditions probably contributes to its stability. The flexibility is also likely to be central to calmodulin's mode of action, as I shall explain below.

The versatility of calmodulin is matched by that of its partner, calcium. The ion is abundant in all biological systems. Its value as a signal is based in part on the fact that the concentration of the ion is between 1,000 and 10,000 times as high in the extracellular fluid as it is inside the cell. Indeed, the cell cannot tolerate a high internal calcium level because the ion combines with essential organic phosphates such as ATP, sequestering them in insoluble salts. The steep steady-state differential between the extracellular and the intracellular concentrations is maintained by the cell membrane, which is ordinarily highly impermeable to the ion, and by mechanisms for disposing of excess calcium; either it is expelled from the cell by an enzymatic pump in the membrane or it is taken up by an intracellular organelle. When the cell is stimulated by an electrical impulse or some other signal, the membrane becomes momentarily permeable to calcium; the influx of the ion that results is detected as a message.

The mechanism by which the calcium ion regulates enzymes such as the calcium-dependent phosphodiesterase can now be described in some detail. The process consists of two successive activations. Phosphodiesterase is essentially inactive without calmodulin and calmodulin is inactive without calcium. When the influx of calcium ions raises their concentration in the cell above a certain threshold, four ions bind to each molecule of calmodulin. (In

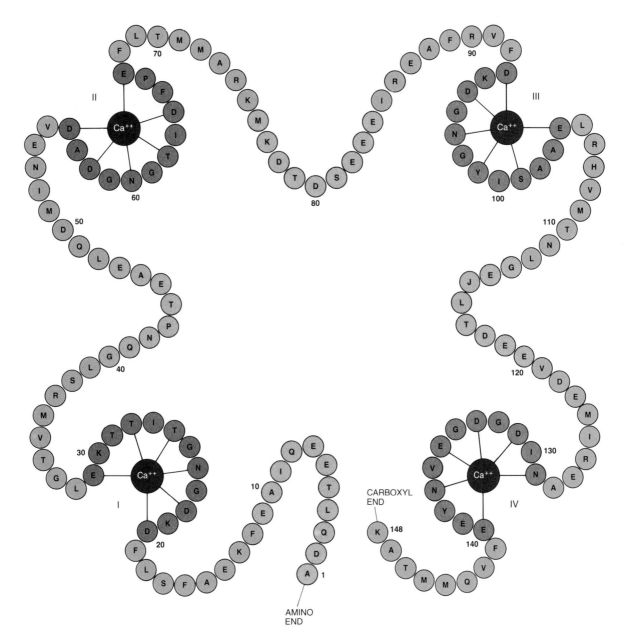

Figure 5.3 CALMODULIN MOLECULE, which mediates many of the regulatory functions of calcium ions, is a single protein chain of 148 amino acid subunits. The chain has four very similar domains, each with a site that binds a calcium ion (Ca++). The schematic diagram of the molecule emphasizes the putative structure of the four calcium-binding sites as proposed by Robert H. Kretsinger of the University of Virginia. Kretsinger based his proposed structure on an X-ray crystallographic study of parvalbumin, a calcium-binding protein with amino acid sequences closely related to those of calmodulin. Each binding site is a loop (*dark color*) flanked by a helical region (*light color*). Each calcium ion is thought to be bonded to six amino acids. The sequence was established by Thomas C. Vanaman and colleagues.

A ALANINE
D ASPARTATE
E GLUTAMATE
F PHENYLALANINE
G GLYCINE
H HISTIDINE
I ISOLEUCINE
J TRIMETHYL LYSINE
K LYSINE
L LEUCINE

M METHIONINE
N ASPARAGINE
P PROLINE
Q GLUTAMINE
R ARGININE
S SERINE
T THREONINE
V VALINE
Y TYROSINE

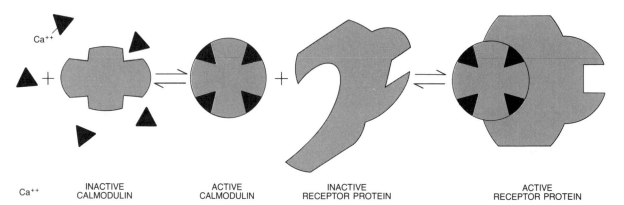

Ca++

| INACTIVE CALMODULIN | ACTIVE CALMODULIN | INACTIVE RECEPTOR PROTEIN | ACTIVE RECEPTOR PROTEIN |

Figure 5.4 MECHANISM by which calmodulin mediates the biological action of calcium ions is depicted in this highly schematic diagram. Neither calcium alone nor calmodulin alone is active. The binding of four calcium ions to calmodulin changes the shape of the protein and thereby activates it. The activated calmodulin is able to interact with an enzyme (or some other protein). The interaction changes the enzyme's shape, with the result that the enzyme is activated. The flexibility of the calmodulin molecule is probably an important factor in process.

the case of some other enzymes the number of ions per calmodulin molecule may be smaller.)

On binding calcium the calmodulin molecule takes on a new, more compact shape and so becomes activated (see Figure 5.4). More specifically, Daniel R. Storm of the University of Washington School of Pharmacology has found that a hydrophobic (water-repellent) region of the molecule is exposed and apparently interacts with the inactive phosphodiesterase (or other enzyme). The enzyme in turn takes on a new conformation and its catalytic activity is enhanced. Any excess of inflowing calcium is taken up by organelles or is pumped out of the cell. The two activation reactions are reversible. When the stimulation of the cell ends, the concentration of calcium ions falls to its steady-state level. The calmodulin molecule releases its ions and returns to its inactive shape, thereby dissociating the complex of calmodulin and phosphodiesterase. The enzyme is thereby rendered inactive and the calcium-initiated reaction ends.

Calmodulin has now been found in all organisms above the level of bacteria and in every kind of cell in such organisms; in other words, it is apparently a component of all eukaryotic (nucleated) cells. Calmodulins from phylogenetically diverse sources have identical or at least very similar biological and biochemical properties; even the sequence of the amino acids along the protein chain appears to have been highly conserved. The result is that the protein lacks both species specificity and

tissue specificity: calmodulin from a unicellular protozoan will function in a test tube to activate an enzyme such as phosphodisterase from the bovine brain.

Recognition of calmodulin's ubiquity and importance came only slowly, as it was found to be present in tissues other than the brain and in association with enzymes other than phosphodiesterase. In the early 1970's Thomas C. Vanaman of the Duke University Medical Center was studying a brain protein that seemed to be a version of troponin C, the calcium receptor in striated muscle. When my colleagues and I published the amino acid composition of calmodulin, Vanaman noticed that it was strikingly similar to the composition of his protein, and he suspected that the two molecules might be related. He sent me a sample of the troponinlike protein to see if it would stimulate phosphodiesterase. It worked about as well as our own calmodulin; indeed, Vanaman's protein was calmodulin. This finding and the results of other studies led to the gradual realization that earlier reports locating troponin C in the brain and in other tissues such as blood platelets were in error. Troponin C is probably found only in cardiac and skeletal muscle; the troponinlike protein in other tissues is calmodulin.

In my laboratory at St. Jude we found that the concentration of calmodulin in various tissues does not correspond to the distribution of phosphodiesterase. This raised the question of whether calmodulin has functions other than the activation of phosphodiesterase. Laurence S. Bradham of the

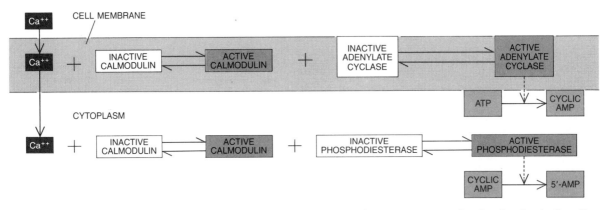

Figure 5.5 SEQUENTIAL ACTIVATION of the enzymes adenylate cyclase and phosphodiesterase in the brain may proceed according to this scheme. When the arrival of a nerve impulse makes the cell's outer membrane permeable to calcium, the ions enter the membrane. They activate calmodulin in the membrane, which in turn activates adenylate cyclase; this enzyme catalyzes the conversion of ATP into the intracellular messenger cyclic AMP. After having fulfilled some message-bearing function in the cell the cyclic AMP is degraded to 5'-AMP by phosphodiesterase. The breakdown enzyme, like adenylate cyclase, is activated by calmodulin. It is thought that calcium ions, having diffused through the cell membrane, bind to cytoplasmic calmodulin, which then activates phosphodiesterase.

University of Tennessee reported in 1972 that brain adenylate cyclase, the enzyme that catalyzes the synthesis of cyclic AMP, requires calcium for maximum activity. Charles O. Brostrom and Donald J. Wolff of the Rutgers Medical School and also Bradham in collaboration with our group went on to do experiments demonstrating that in the brain calmodulin activates adenylate cyclase as well as phosphodiesterase (see Figure 5.5).

The regulation of the synthesizing enzyme and that of the breakdown enzyme by calmodulin seem to be nicely complementary. Adenylate cyclase is associated with the cell membrane, whereas the calcium-dependent phosphodiesterase is in the cytoplasm. As inflowing calcium passes through the outer membrane it binds to calmodulin molecules in the membrane and thus activates adenylate cyclase; the enzyme converts ATP into cyclic AMP. Somewhat later the calcium ions reach and bind to calmodulin molecules in the cytoplasm and thus activate phosphodiesterase. The sequential activation of the two enzymes may be responsible for the transient increase in intracellular cyclic AMP that is commonly observed when certain tissues are stimulated.

James D. Porter of the University of Cincinnati College of Medicine recently showed that the two enzymes differ in their sensitivity to calcium ions. This could mean that during the initial phase of the calcium influx rather low levels of the ion suffice to activate adenylate cyclase. As the calcium level rises the ion does two things: it activates phosphodiesterase and it also somehow inhibits adenylate cyclase, thereby quickly reducing cyclic AMP to its preactivation steady-state level.

The finding that calmodulin regulates a second enzyme in the brain still could not explain the high concentration of calmodulin there (far in excess of what is needed to regulate adenylate cyclase and phosphodiesterase), to say nothing of the presence of calmodulin in tissues lacking either of the enzymes. Soon the search was on for other functions. The list is still growing (see Figure 5.6), and most of us in the field do not expect the pace of discovery to slacken for several years.

In 1973 Guy H. Bond of the Medical College of Virginia found that in human red blood cells there is a factor that stimulates calcium-ATPase, the enzymatic pump in the cell membrane that moves calcium out of the cell. Frank F. Vincenzi of the University of Washington School of Medicine and John T. Penniston of the Mayo Clinic identified the pump activator as calmodulin, which has since been found to stimulate calcium-ATPase in other cells as well. Calmodulin, then, has a double set of functions: it not only transmits calcium's message to receptor enzymes but also modulates the intracellular con-

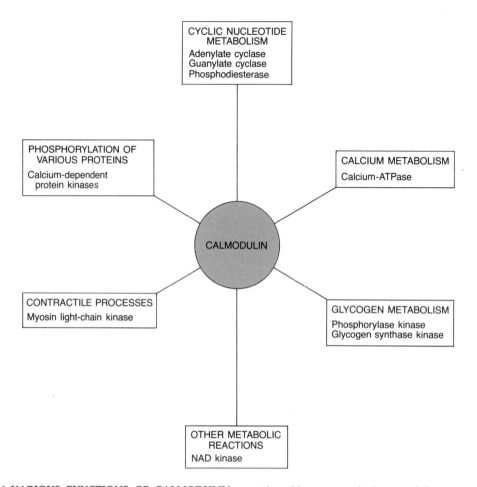

Figure 5.6 VARIOUS FUNCTIONS OF CALMODULIN are summarized, together with some of the enzymes known to serve each function. Adenylate cyclase and guanylate cyclase catalyze the synthesis of cyclic AMP and cyclic GMP respectively; phosphodiesterase breaks down both cyclic molecules. Calcium-ATPase expels calcium ions from the cell. Phosphorylase kinase and glycogen synthase kinase respectively control the enzymes that initiate the breakdown and the synthesis of glycogen. NAD kinase synthesizes NADP from the cofactor NAD. Myosin light-chain kinase controls contraction in smooth-muscle and nonmuscle cells. There appear to be other calcium-dependent protein kinases that phosphorylate various proteins.

centration of the ion (something I had in mind when I coined the name calmodulin). Having sensed the arrival of calcium and thereby been activated to stimulate an enzyme, calmodulin proceeds to turn on the pump that rids the cell of unneeded calcium.

As I have mentioned, the first calcium receptor to be identified was troponin C in striated muscle. The arrival of a nerve impulse at the muscle releases calcium from storage in the part of the muscle cell called the sarcoplasmic reticulum (see Figure 5.7).

The ion binds to troponin C and alters the shape of the troponin molecule, initiating a series of interactions among muscle proteins that eventually catalyzes the hydrolysis of ATP to release energy for muscle contraction. In the contraction of smooth muscle and of filaments in nonmuscle cells, however, calcium's effect is mediated by calmodulin rather than troponin. David J. Hartshorne of the University of Arizona and others have shown that calmodulin, having bound calcium ions, activates a protein kinase: an enzyme that phosphorylates

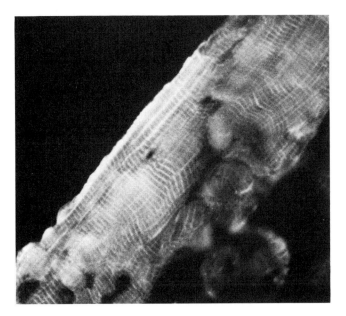

Figure 5.7 IN SKELETAL MUSCLE calmodulin is apparently associated with one band (thought to be the one designated *I*) **in the repeated pattern of transverse striations characteristic of such muscle cells. Calmodulin is also found in a longitudinal structure, probably the sarcoplasmic reticulum. In this micrograph a segment of the gastrocnemius muscle in the leg of a rat is enlarged 800 diameters. The location of antibody that binds to calmodulin is revealed by the binding of a second antibody labeled with a fluorescent dye.**

(adds a phosphate group to) another enzyme or some other protein. The kinase activated by calmodulin in smooth muscle phosphorylates a regulatory component of the protein myosin, thereby initiating interactions that presumably are similar to those controlled by troponin C in striated muscle.

Muscle contraction, which consumes energy, is metabolically coordinated with a sequence of events that supplies energy. Glycogen, the storage form of glucose, is the main source of quickly available energy, and large reserves of glycogen are close at hand in muscle cells. Because glycogen is a polymer of glucose (a long chain of linked glucose molecules), a necessary step toward supplying energy is the depolymerization of glycogen. When striated muscle is stimulated, calcium ions released from the sarcoplasmic reticulum trigger not only contraction but also the depolymerization of glycogen. They do so by binding not to free calmodulin molecules but to a calmodulin unit that is itself a subunit of the enzyme being activated: a kinase whose structure has been worked out by Philip Cohen of the University of Dundee. The activated kinase adds phosphate groups to the enzyme phosphorylase, converting it from the inactive *b* form into the active *a* form. Phosphorylase *a* initiates a breakdown of glycogen, making glucose available to generate ATP, the universal energy transducer in cells (see Figure 5.8).

There is more to the story. Muscle cells have an enzyme for storing glucose as well as one for bringing it out of storage. The storage enzyme, glycogen synthase, initiates a reaction that links glucose units to form glycogen, and its polymerizing activity needs to be turned off when glucose is required by the muscle cell. In this case phosphorylation turns the enzyme off rather than on. Thomas R. Soderling of the Vanderbilt University School of Medicine recently showed that calmodulin activates glycogen synthase kinase, an enzyme that phosphorylates and thereby inactivates glycogen synthase. Thus calcium and calmodulin coordinate the suppression of glycogen synthesis with the stimulation of glycogen breakdown. Both actions are linked by calcium (by way of calmodulin) to the contractile event they serve.

The phosphorylation of a protein by a kinase is often a key step in a regulatory process. There is evidence that in addition to the three protein kinases implicated in smooth-muscle contraction and in glycogen metabolism there may be other calmodulin-dependent kinases. Paul Greengard of the Yale University School of Medicine has detected one such kinase that seems to phosphorylate a number of different proteins depending on the tissue in which it is active. Robert DeLorenzo of Yale has found in the membrane of nerve-cell terminals

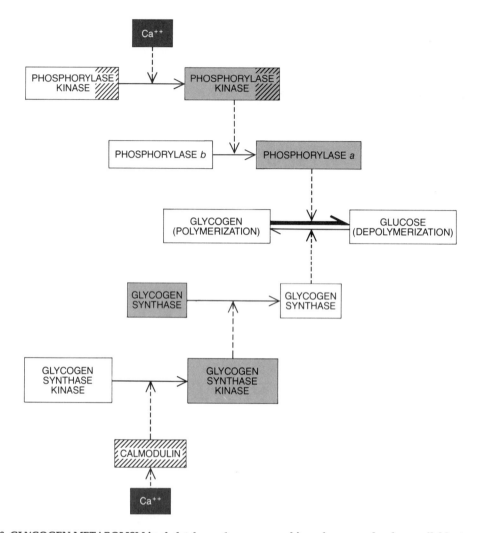

Figure 5.8 GLYCOGEN METABOLISM in skeletal muscle is regulated by calcium ions, which bind to a calmodulin molecule (*hatching*) that is a subunit of the enzyme phosphorylase kinase. The activated kinase (*color*) adds phosphate groups to the inactive enzyme phosphorylase *b*, converting it into the active form, phosphorylase *a*. The activated phosphorylase initiates the breakdown of glycogen, making glucose molecules available to supply the energy required for muscle contraction. Concurrently the polymerization of glucose to form glycogen is stopped. Calcium ions bind to calmodulin, which activates the enzyme glycogen synthase kinase. The kinase adds phosphate groups to glycogen synthase. In this case the phosphorylation inactivates the enzyme.

a calmodulin-dependent kinase that phosphorylates the protein tubulin. The phosphorylation changes the physical and chemical properties of the tubulin, which then aggregates to form filamentous structures. (The filaments do not seem to be microtubules, the cablelike structures usually formed by tubulin.) DeLorenzo thinks the filaments may interact with the membrane to facilitate the release of norepinephrine. Norepinephrine is a neurotransmitter, one of the substances that relay a nerve impulse from one nerve cell to another one or to a muscle cell.

Calmodulin may also have a role in the synthesis of neurotransmitters. The catecholamine transmitters dopamine, norepinephrine and epinephrine are synthesized from the amino acid tyrosine in several steps, the first of which is catalyzed by the enzyme tyrosine 3-monooxygenase. Hitoshi Fujisawa of the Asahikawa Medical College has shown that the enzyme's activity depends on a phosphorylation by a calmodulin-dependent kinase. Tryptophan 5-monooxygenase, which catalyzes the conversion of the amino acid tryptophan into the neurotransmitter serotonin, is similarly controlled by a calmodulin-dependent phosphorylation.

The cofactor called nicotinamide adenine dinucleotide phosphate (NADP) is required for the synthesis of many cellular constituents, including steroids, nucleotides (the subunits of nucleic acids) and fatty acids. NADP can be synthesized from nicotiamide adenine dinucleotide (NAD) by a specific kinase, which Milton J. Cormier of the University of Georgia has found is dependent on calmodulin. NAD and NADP are required for many cellular activities during the early development of the embryo. Calcium is known to be important for fertilization. David Epel of the Hopkins Marine Station of Stanford University and Robert W. Wallace and I have shown than an increase in the internal calcium level soon after fertilization of the sea-urchin egg leads to a transient activation of the NAD kinase by calmodulin. The result is increased synthesis of NADP, which could contribute to metabolic processes responsible for the successive cell divisions that produce the early embryo.

Cyclic guanosine monophosphate (cyclic GMP) is a nucleotide similar to cyclic AMP; like the latter molecule it is thought to serve as an intracellular messenger, although its functions are not well understood. Recently Yoshio Watanabe of the University of Tsukuba has reported that in the protozoan *Tetrahymena* calmodulin activates guanylate cyclase, the enzyme that catalyzes the synthesis of cyclic GMP. Whereas calmodulin from any species or tissue usually functions in other tissues or species, the protozoan guanylate cyclase seems to be activated only by its own calmodulin. Mammalian guanylate cyclase is known to require calcium for maximum activity, but whether the ion's effect on the mammalian enzyme is mediated by calmodulin remains to be shown.

I have discussed some of the processes in which calmodulin regulates a specific enzyme whose function is known. The protein undoubtedly has other activities for which there is so far only suggestive evidence. It has been found in many tissues where its mode of action is not yet known. In other cases proteins have been identified that clearly bind to calmodulin but whose own function is not yet established.

One such protein is calcineurin, which is present, along with calmodulin, in many regions of the mouse brain. Calcineurin has two subunits, one of them about the size of calmodulin and the other

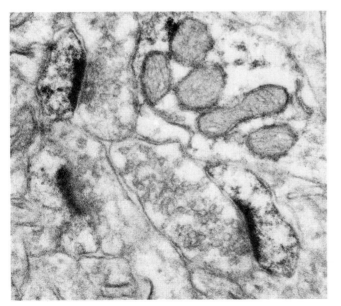

Figure 5.9 IN THE BRAIN calmodulin seems to be found near the postsynaptic region of a nerve cell, where impulses transmitted from another cell are received. In this micrograph, a thin section from the basal ganglia region of rat brain is enlarged 70,000 diameters. The section was incubated first with a rabbit antibody to calmodulin and then with an antibody, labeled with the enzyme peroxidase, to the rabbit antibody. A stain precipitated by the peroxidase shows that calmodulin has become concentrated at the postsynaptic densities.

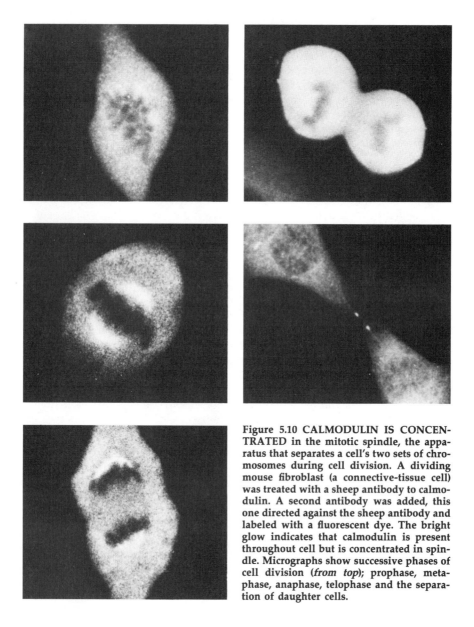

Figure 5.10 CALMODULIN IS CONCEN-
TRATED in the mitotic spindle, the apparatus that separates a cell's two sets of chromosomes during cell division. A dividing
mouse fibroblast (a connective-tissue cell)
was treated with a sheep antibody to calmodulin. A second antibody was added, this
one directed against the sheep antibody and
labeled with a fluorescent dye. The bright
glow indicates that calmodulin is present
throughout cell but is concentrated in spindle. Micrographs show successive phases of
cell division (*from top*); prophase, meta-
phase, anaphase, telophase and the separation of daughter cells.

larger. Claude B. Klee of the National Cancer Institute has found that the small subunit, like calmodulin, binds four calcium ions and that the large subunit binds to calmodulin. Calcineurin's function is
not fully understood, but preliminary results from
Cohen's laboratory and mine suggest it is a protein
phosphatase: an enzyme that catalyzes the removal
of a phosphate group from proteins. It is concentrated at postsynaptic densities (see Figure 5.9), near
where the receptors for neurotransmitters are situated, and my graduate student E. Ann Tallant has
shown that its level increases with the formation of

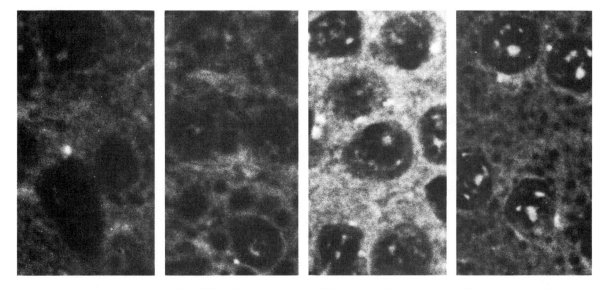

Figure 5.11 HORMONAL STIMULATION of the rat adrenal cortex seems to cause calmodulin to become associated with the nuclei of cortical cells. The micrographs show cortical tissue from adrenal glands removed before the injection of corticotropin (ACTH), which stimulates the cortex to synthesize steroid hormones, and at intervals after the injection. In unstimulated cells (*left*) some calmo-dulin is revealed by immunofluorescence in the cytoplasm. After half an hour (*second from left*) the nuclei show some fluorescence. After an hour (*third from left*) the concentration of calmodulin seems to have increased. After 11 hours (*right*) most of the calmodulin is associated with nuclei, suggesting that calmodulin has a role in some nuclear function of ACTH.

synapses between nerve cells during development. These findings suggest that calcineurin may have some role in neurotransmission.

The metabolism and functions of calcium and cyclic AMP are intertwined. I have explained how in the brain the synthesis and the degradation of cyclic AMP are both regulated by calcium ions by way of calmodulin; the same may be true in some other tissues. Cyclic AMP promotes the uptake of calcium into organelles such as the sarcoplasmic reticulum, a process tending to turn off a calcium-initiated action. There are instances where calcium and cyclic AMP act in opposition. For example, in smooth muscle they seem to have opposite effects on myosin light-chain kinase, the enzyme that initiates contraction. Calcium-activated calmodulin stimulates the enzyme; Robert S. Adelstein of the National Heart, Lung, and Blood Institute has preliminary evidence suggesting that cyclic AMP makes the enzyme refractory to such stimulation, lessening calcium's effect. In still other cases calcium has a function in one tissue that is performed in another tissue by cyclic AMP. The stimulation of

phosphorylase to degrade glycogen is generally accomplished by calcium and calmodulin in muscle and by cyclic AMP in the liver.

The calcium-calmodulin system and the cyclic-AMP system have different attributes, which may explain why the systems are active under different circumstances. Calmodulin is present in every cell (see Figure 5.10) and calcium is plentiful in the extracellular fluid; neither substance needs to be synthesized to initiate an action, and so the calcium system is inherently fast-acting. Cyclic AMP, on the other hand, is essentially synthesized de novo by adenylate cyclase when the latter is stimulated, and that takes a certain amount of time.

The most important receptor for cyclic AMP in eukaryotic cells is a protein kinase that is activated by cyclic AMP and can thereupon phosphorylate a number of different substrate proteins. The calcium system, in contrast, has at least three different calmodulin-dependent kinases that phosphorylate specific substrates, and it probably also has a number of other calmodulin-dependent kinases that phosphorylate various enzymes. Moreover, calcium, through calmodulin, regulates many other

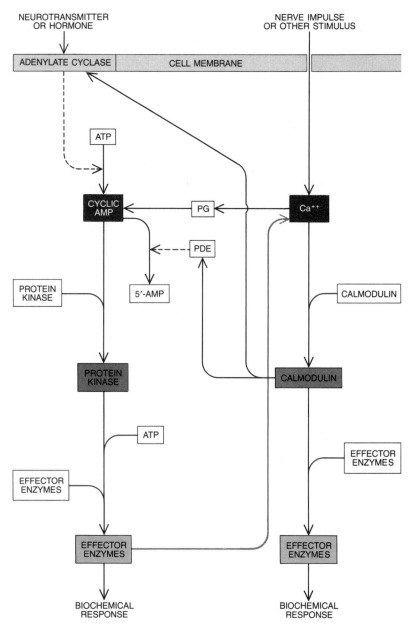

Figure 5.12 INTERRELATION of the cyclic-AMP messenger systems (*left*) and the calcium-calmodulin system (*right*). The arrival of an intercellular message activates adenylate cyclase, which synthesizes cyclic AMP; this messenger activates a protein kinase, which phosphorylates some effector protein, either stimulating or inhibiting its biological activity. Calmodulin, when it is activated by calcium ions, activates an effector protein to initiate a biological response. Calcium stimulates the activity of adenylate cyclase and phosphodiesterase (*PDE*), which breaks down cyclic AMP. Calcium modulates the metabolism of prostaglandin (*PG*), which in turn influences adenylate cyclase metabolism.

enzymes in addition to protein kinases. Because calmodulin is flexible it can interact with a variety of effector enzymes, making the calcium ion much more versatile than cyclic AMP.

In mammals the important cellular messengers are hormones, cyclic AMP (and other cyclic nucleo-

tides) and calcium. For the most part hormones make it possible for cells to communicate with one another (see Figure 5.11); cyclic AMP and calcium carry messages from one part of a cell to another. The three messengers have complementary roles with respect to time as well as distance. For hormones the response time and the duration of action range from minutes to hours and even weeks. For cyclic AMP the range is from seconds to minutes and for calcium it is probably in the millisecond range. Hormones are often designated the "first messengers" and cyclic AMP a "second messenger." Considering the extent to which the metabolism and functions of the calcium ion and cyclic AMP are interconnected and therefore overlap in time and space, it is probably more accurate to refer to both of them simply as cellular messengers or regulators, rather than attempting to order them as being second or third messengers in any particular process (see Figure 5.12).

The discovery of calmodulin has greatly clarified how calcium acts at the molecular level. In elucidating the various roles of calmodulin an agent that counteracts the activity of the protein has been particularly helpful. Benjamin Weiss of the Medical College of Pennsylvania reported in the mid-1970's that trifluoperazine, an antipsychotic agent, inhibits the activity of calmodulin by preventing the protein's interaction with receptor enzymes. The drug has aided immensely in the identification of some of the biological functions of calmodulin. Moreover, it has stimulated efforts to design and identify drugs that affect either the activities of calcium mediated by calmodulin or other aspects of calcium's action and metabolism.

How is it that calmodulin, the ever-present participant in calcium-initiated cellular processes, remained hidden from investigators through several decades during which those processes were under intensive study? The answer lies primarily in calmodulin's ubiquity and abundance. Unlike cyclic AMP, for example, calmodulin seems not to be the limiting factor in a cellular activity; since it is always present, both in cells and in cell extracts, it is never missed. Its role was disclosed only by its inadvertent removal from phosphodiesterase during the routine purification of the enzyme. The loss of enzyme activity in the course of purification is usually just an annoyance. In the case of phosphodiesterase I was fortunate in that the discrepancy between the results of two assay procedures pointed to the nature of the problem: the removal of an activator, which turned out to be calmodulin.

The Cycling of Calcium as an Intracellular Messenger

A generalized increase in the concentration of calcium in a cell has usually been portrayed as a switch turning cellular processes on and off. But the ion's role in prolonged responses belies this traditional model. Calcium appears to operate as a signal within a specially restricted intracellular domain.

. . .

Howard Rasmussen
October, 1989

Two of the more remarkable events in the course of evolution were the development of the exoskeleton of mollusks and, hundreds of millions of years later, the bony endoskeleton of higher animals. Each development represented a new biological use for calcium. Calcium salts in the form of shell, bone and tooth are familiar materials of bioarchitecture; they are visible signs of the importance of calcium in the growth and function of organisms.

There is less general awareness, however, of an older and more pervasive role of the calcium ion: within a wide variety of animal cells, calcium serves as an almost universal ionic messenger, conveying signals received at the cell surface to the inside of the cell. The calcium ion is involved in such diverse processes as the regulation of muscle contraction, the secretion of hormones, digestive enzymes and neurotransmitters, the transport of salt and water across the intestinal lining and the control of glycogen metabolism in the liver.

Whereas the calcification that produces bone in-

volves the ordered deposition of large amounts of salts, the intracellular messenger function involves minute flows of calcium ions across the membranes of cells. In fact, calcium ions can carry out their informational role only at very low and tightly controlled concentrations, because higher ones are detrimental to normal cell function.

Cells have a simple but elegant set of mechanisms by which they regulate intracellular calcium levels. The mechanisms work mainly by controlling the movement of calcium ions across three membranes: the plasma membrane, which surrounds the cell; the inner membrane of the mitochondrion, a cell's energy-producing organelle; and the membranes of compartments that contain reserves of calcium ions, which are called the sarcoplasmic reticulum in muscle cells and calcisomes in nonmuscle cells. Although the concentration of calcium inside a cell remains fairly constant, the flow across the plasma membrane (the amount of influx and efflux) can vary significantly.

Recently it has become clear that such calcium-ion "cycling" across the plasma membrane is part of a complex chain of events by which cells generate sustained responses to stimuli in their environment. Calcium's role in sustained cellular responses, such as the secretion of insulin or the contraction of the smooth muscle surrounding the blood vessels, has historically been more elusive than its part in transient responses, such as the contraction of skeletal muscle. My colleagues and I have been able to piece together a picture of how cycling across the plasma membrane mediates sustained cellular responses. We have already found similar mechanisms operating in three disparate cell systems. Our findings have led us to a more sophisticated understanding of calcium as an intracellular messenger than we had even five years ago.

The sensitivity of a cell to very small changes in calcium concentration reflects the ion's very low concentration within the cell. The concentration of calcium ions is generally 10,000 times greater in the fluid surrounding the cell than in the intracellular fluid, or cytosol. Maintenance of the concentration difference depends on two features of the plasma membrane: its low permeability to calcium and the presence of membrane-bound "pumps" that drive calcium out of the cell, against the concentration gradient. At resting conditions the rate of calcium-ion leakage or influx into the cytosol is balanced by a similar rate of pump-driven efflux of calcium ions.

The Classic Calcium Signal

When a cell is stimulated by an extracellular signal, channels in its plasma membrane open and allow calcium ions to enter at from two to four times the normal rate. Such channels allow calcium but no other ions to flow into the cell cytosol. Some channels open when a neurotransmitter changes the voltage difference that normally exists across the cell membrane; others open when a hormone or neurotransmitter interacts with a cell-surface receptor that is inked to the channels.

The traditional view of calcium as an intracellular messenger is fairly straightforward [see "The Calcium Signal," by Ernesto Carafoli and John T. Penniston; Scientific American, November, 1985; Offprint 1564]. Stimulation by a hormone or a neurotransmitter increases the calcium-ion concentration in the cytosol as calcium channels open in the plasma membrane or as calcium is released from the sarcoplasmic reticulum or from calcisomes. When the concentration rises, calcium-binding proteins in the cytosol, such as the specific receptor calmodulin, attach to calcium ions; the calcium-protein complexes then interact with other proteins in the cell to alter their functions. When the calcium concentration in the cytosol falls again, the ions dissociate from the receptor proteins and the system turns off.

In this scenario, calcium acts as a simple on-off switch that conveys information from the cell surface to the cell interior. Calcium does in fact serve as such a switch in several transient cellular responses, including the secretion of neurotransmitters by nerve cells and the contraction of skeletal and cardiac muscle cells. In each case, the rise in calcium-ion concentration in the cytosol initiates the response, and the fall in calcium-ion concentration terminates it.

A First Order of Complexity

While the preceding model of calcium's messenger action was being developed, a similar model was proposed for another messenger molecule known as cyclic adenosine monophosphate (cyclic AMP, or cAMP). It was thought that the synthesis of cAMP at the plasma membrane and its breakdown in the cytosol represented an on-off switch much like variations in calcium-ion concentration. At first the two switches were thought to operate independently.

Now, however, biologists believe that cAMP and calcium usually work together to regulate cell behavior. For example, cAMP can control the rate of calcium cycling across the plasma membrane, and calcium can regulate the enzymes responsible for the synthesis and destruction of cAMP. A single hormone acting by way of a single receptor can cause a simultaneous increase in calcium-ion influx and cAMP production.

Finally, both calcium and cAMP exert many of their cellular effects by controlling the activity of a particular class of enzymes called protein kinases. Protein kinases catalyze the transfer of phosphate groups from a molecule called adenosine triphosphate (ATP) to other proteins. The addition of a phosphate group alters protein function; indeed, widespread protein phosphorylation is thought to underlie the changes in cell behavior induced by some extracellular signals.

The picture of the individual messenger functions of calcium and cAMP has changed as well. The view that calcium ions and cAMP serve as simple switches is not borne out in all contexts. In particular, that paradigm cannot account for sustained cellular responses to the sustained presence of an extracellular messenger.

My colleagues and I have focused on the role of the calcium ion in sustained cellular responses. We have investigated three such responses: the secretion of the hormone aldosterone, which regulates potassium metabolism, by the cells of the adrenal glomerulosa (part of the adrenal gland); the secretion of insulin by the beta cells of the islets of Langerhans in the pancreas; and the contraction of the smooth muscle cells surrounding the trachea and many blood vessels. In spite of marked differences in the nature of these responses and the extracellular signals that elicit them, we have found that the calcium ion carries out its role as messenger in much the same manner in all three cell types.

A Second Order of Complexity

Until several years ago, biologists thought that sustained, calcium-dependent cellular responses resulted from sustained rather than transient increases in the concentration of calcium ions in the cytosol. This idea was not based on direct measurements of calcium. Instead it was founded on the belief that calcium acted in a manner similar to cAMP, which undergoes a sustained increase in concentration in response to an appropriate stimulus. Yet when the cytosolic concentration of calcium ions was actually measured in stimulated cells, it was found that although the calcium-ion concentration rose as predicted, it did so only transiently and fell back to its base level within a few minutes when the cells continued to respond for hours (see Figure 6.1).

This paradox forced us to reconsider the classic view of calcium-messenger function. Accordingly, my colleagues Itaru Kojima, Kumiko Kojima, William J. Apfeldorf and Paula Q. Barrett further analyzed the alterations in calcium-ion metabolism induced in the cells of the adrenal glomerulosa by the hormone angiotensin II—the trigger for aldosterone secretion. We learned that whereas in such cells angiotensin II does cause only a transient rise in the cytosolic calcium-ion concentration, it also causes a sustained twofold increase in the influx of calcium ions.

This finding presented a second paradox. It was generally thought that a sustained increase in calcium influx would give rise to a sustained increase in the cytosolic calcium-ion concentration, but clearly that was not the case. It was also assumed that a sustained increase in calcium influx would lead to an increase in the total calcium in a cell, but we found no such increase. We therefore concluded that during the sustained phase of the adrenal-cell response, angiotensin II causes a sustained increase in calcium cycling across the plasma membrane.

The molecular basis for this remarkable ability of the plasma membrane resides in the properties of the calcium-ion pump. It turns out that this pump is activated by a complex of calmodulin and calcium (see Figure 6.2). When cytosolic calcium levels rise, the complex interacts with the pump to enhance both its efficiency and its sensitivity to calcium ions. The efficiency of the pump is further enhanced when it is phosphorylated by a calcium-activated protein kinase called protein kinase C (PKC). The stimulation of the pump by the calcium-calmodulin complex and calcium-activated phosphorylation enables calcium efflux to compensate for the increased influx.

Having discovered the change in the calcium-cycling rate and defined the mechanisms by which it occurs, my colleagues and I explored the possibility that the cycling serves as a messenger during the sustained cellular response. We found that if we blocked the increase in calcium cycling, the response was transient rather than sustained. In other words, calcium cycling is critical to maintaining a sustained response. We concluded that such cycling acts as a messenger because it leads to a change in the calcium-ion concentration in a restricted part of the cell—the "submembrane" domain within or just beneath the plasma membrane.

The next question we wanted to answer was just how this messenger acts. We found that simply increasing calcium influx was not enough to induce a response. Thus, an increase in calcium cycling is a necessary but not sufficient condition for the induction of a sustained response. It became clear that a calcium-sensitive, plasma membrane-associated "transducer" is also required to read the calcium-ion message and convert it into a form that can affect the rest of the cell. Several of these transducer-molecules have been discovered in adrenal cells and other cell types, but the one of present interest is protein kinase C—coincidentally, the same enzyme that regulates the activity of the calcium-ion pump.

The activation of PKC, which enables it to act as a transducer for the sustained, submembrane calcium signal, is linked to the turnover of a class of molecules called inositol phospholipids. The metabolism

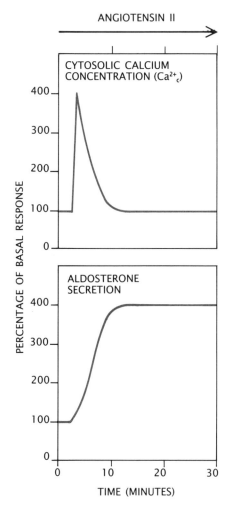

ANGIOTENSIN II

CYTOSOLIC CALCIUM CONCENTRATION (Ca$^{2+}_c$)

ALDOSTERONE SECRETION

PERCENTAGE OF BASAL RESPONSE

TIME (MINUTES)

Figure 6.1 DISSOCIATION with time of changes in calcium-ion concentration (*top*) and cell response (*bottom*) is not predicted by the classic model of calcium-messenger action. These data describe the secretion of the steroid hormone aldosterone from cells of the adrenal glomerulosa (part of the adrenal gland) in response to the hormone angiotensin II. The concentration of calcium in the cytosol, or intracellular fluid, spikes one minute after angiotensin II is added, but aldosterone secretion lasts for more than 30 minutes.

of these molecules is regulated by a certain class of hormones and neurotransmitters (see Chapter 4, "The Molecular Basis of Communication within the Cell," by Michael J. Berridge). When such an agent binds to its receptor, an enzyme linked to the receptor catalyzes the breakdown of phosphatidylinositol 4,5-biphosphate (PIP$_2$), which is a component of the cell membrane. The reaction yields inositol-1,4,5-triphosphate (IP$_3$) and diacylglycerol (DAG, or DG). Released into the cytosol, IP$_3$ induces the liberation of calcium ions from the calcisomes. Calcium ions released by this mechanism cause a transient increase in the calcium concentration in the cytosol. The rise promotes the formation of calcium-calmo-

dulin complexes, which go on to assist in the phosphorylation of a specific subset of proteins by activating certain protein kinases.

A Two-Pronged Response

The calcium transient and DAG, the other product of PIP$_2$ breakdown, together cause PKC to associate with the plasma membrane. Unlike IP$_3$, DAG remains in the membrane; as long as the DAG content of the membrane stays high, PKC remains associated with the membrane as well. The transient release of calcium ions from the calcisomes and the migration of PKC from the cytosol to the plasma

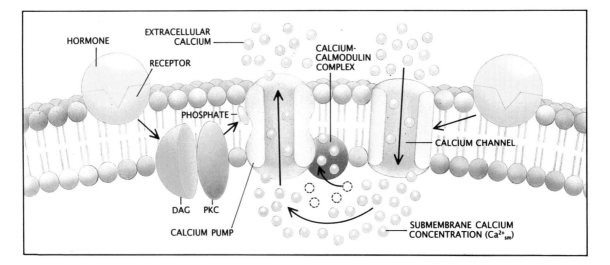

Figure 6.2 AUTOREGULATION of the calcium flowing across the plasma membrane is achieved by the control of calcium pumps. The entry of calcium through membrane channels (*right*) increases when a hormone interacts with its receptor. The rise in submembrane calcium concentration (Ca^{2+}_{sm}) stimulates the activity of a calcium pump (*left*) by activating the calcium-binding protein calmodulin and the enzyme protein kinase C (PKC). In this way calcium efflux balances calcium influx. The rise in submembrane calcium caused by such "cycling" constitutes a new kind of calcium messenger.

membrane are the hallmarks of the initial stage of a sustained cellular response to an extracellular signal.

Receptor activation also causes the twofold increase in calcium-ion influx seen at the plasma membrane. The mechanism by which this increase occurs and the type of calcium channel that opens vary from one cell type to the next. It is not yet clear whether the calcium influx increases as a direct result of receptor activation or as a result of a signal generated by PIP_2 hydrolysis. It is clear, however, that the increase in calcium influx, and therefore in calcium cycling, is a critically important messenger during the sustained phase of the response.

It is thought that the cytosolic form of PKC is relatively inactive. When PKC associates with the plasma membrane, however, it comes in contact with phospholipids (the major components of the membrane) that increase the enzyme's maximal rate of activity by a factor of from 25 to 30 and its sensitivity to calcium ions by a factor of 100 or more. This calcium-sensitive, plasma membrane-associated form of PKC is what acts as a transducer during the second phase of the sustained cellular response. It is the target of the localized change in the concentration of calcium ions brought about by

increased calcium cycling. The increase in the submembrane concentration of calcium ions somehow increases the rate at which PKC helps to phosphorylate other proteins.

The key feature of the model that has emerged from our studies of a sustained response in adrenal cells is the operation of two temporally distinct branches of the calcium-messenger system: a calmodulin branch, active during the initial phase of response, in which the transient, IP_3-induced rise in the cytosolic concentration of calcium acts on calmodulin-dependent protein kinases to alter the phosphorylation of one subset of cellular proteins; and a PKC branch, in which the rise in calcium concentration in the submembrane domains acts on plasma membrane-associated PKC to alter the phosphorylation of a different subset of cellular proteins involved in mediating the second, sustained phase of the cellular response (see Figure 6.3).

This two-branch model of calcium-messenger action seems to account not just for the secretion of aldosterone from adrenal cells in response to angiotensin II but also for the stimulation of insulin secretion from beta cells and the contraction of smooth muscle. In those systems, however, the

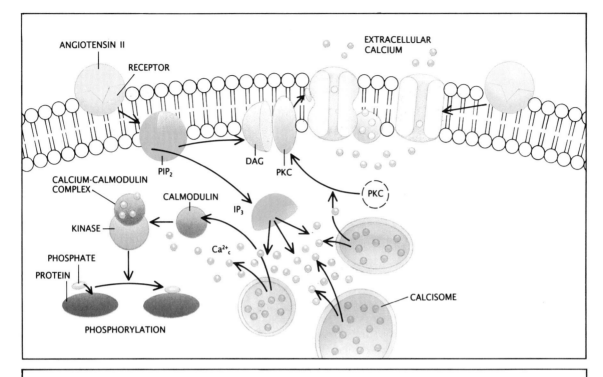

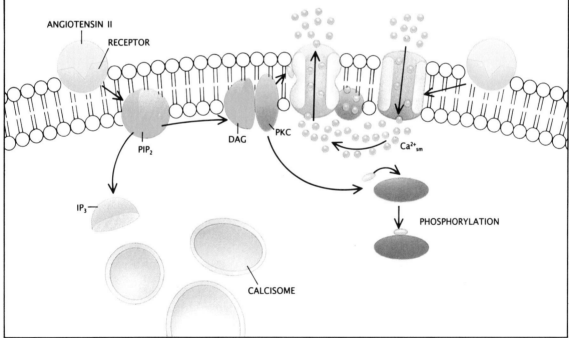

Figure 6.3 ACTIVATION of an adrenal glomerulosa cell by angiotensin II illustrates how calcium acts as a messenger in two different ways. In the initial phase (*top*), the binding of an extracellular signal to its receptor prompts the breakdown of PIP_2 into IP_3 and DAG. IP_3 causes the release of calcium ions from calcisomes, resulting in a transient rise in cytosolic calcium (Ca^{2+}_c). The ions bind to calmodulin, and the calcium-calmodulin complex activates protein kinases. The phosphorylated proteins initiate the cellular response, the secretion of aldosterone. In the sustained phase (*bottom*, angiotensin II increases calcium cycling across the membrane, and the rise in the submembrane concentration of calcium (Ca^{2+}_{sm}) activates the membrane-associated PKC, which brings about the phosphorylation of different proteins that sustain aldosterone secretion.

operation of the calcium-messenger system depends heavily on the activity of the cAMP-messenger system.

Additional Complexity

The interactions between calcium and cAMP in the regulation of insulin secretion are too complex to be addressed fully in this chapter. I shall limit my discussion to the action of acetylcholine, a neurotransmitter that attaches to specific receptors on beta cells and thereby brings about in those cells the same signaling events that angiotensin II evokes in adrenal cells. In beta cells, however, the situation is more complex than it is in adrenal cells, because both the intracellular content of cAMP and the extracellular concentration of glucose determine how effective the signals acetylcholine generates will be.

The intracellular content of cAMP can be increased by the binding of certain hormones to specific receptors on the beta cell. One such hormone, called gastric inhibitory peptide (GIP), is released from intestinal mucosal cells in the course of food intake and digestion. When my colleagues Walter S. and Kathleen C. Zawalich and I studied the combined effects of acetylcholine and GIP on insulin secretion from isolated islets in the laboratory, we found that, as we have anticipated, acetylcholine stimulates the breakdown of PIP_2, and GIP boosts the production of cAMP. The remarkable observation was that the effect of such stimulation on insulin secretion depended on the extracellular glucose concentration. For example, when the glucose concentration was similar to that present in the blood before a meal, combined acetylcholine and GIP increased insulin secretion only briefly and to a small extent. If, however, the islets were exposed to a glucose concentration about 50 percent higher, which is similar to the one present in the blood from 30 to 60 minutes after a meal, the same combination of acetylcholine and GIP caused a significant and sustained increase in insulin secretion.

The results from these studies and from studies in other laboratories led us to conclude that the different concentrations of glucose altered the effects of acetylcholine and GIP on calcium influx. In the case of the lower glucose concentration, acetylcholine and GIP had little or no effect on calcium influx, but at higher glucose concentrations, they stimulated the influx by way of a specific type of voltage-dependent membrane channel. In the beta cell as in the adrenal cell, a sustained increase in calcium in-

flux (and hence calcium cycling) is essential for a sustained cellular response.

Glucose acts as a conditional modifier of beta-cell responsiveness by controlling the electrical potential of beta-cell membranes. When blood glucose is low, the membrane potential is high, and voltage-dependent calcium channels in the cell membrane stay closed even when acetylcholine initiates a depolarization—a reduction in voltage across the membrane. When blood glucose is higher, however, the membrane becomes partially depolarized: it is poised so that the additional depolarization caused by acetycholine induces channels to open, letting calcium into the cell.

Furthermore, the increase in cAMP concentration that results from stimulation by GIP also causes certain latent calcium channels to become sensitive to voltage, so that an increasing number of channels respond when an appropriate change in membrane potential occurs. Thus, the rate of calcium-ion influx—and hence of calcium cycling—is increased by two different mechanisms when acetylcholine and GIP act in concert at the proper glucose concentration.

From a physiologic point of view, the control of the calcium-ion influx rate in this way provides a fail-safe mechanism to prevent an inappropriate secretion of insulin when the blood glucose is low—before a meal, for example. A similar fail-safe system operates in adrenal glomerulosa cells. A major effect of the aldosterone they secrete is to lower the concentration of potassium in the blood. A rise in extracellular potassium-ion concentration partially depolarizes the adrenal-cell plasma membrane; the system is thereby poised so that a particular type of voltage-dependent channel opens when angiotensin II is present. When blood potassium falls to a low value, the plasma membrane becomes hyperpolarized, the channels are closed and angiotensin II cannot open them. Consequently, angiotensin II cannot elicit a sustained increase in calcium influx or in aldosterone secretion when blood potassium is low. This fail-safe mechanism prevents secretion of aldosterone when an increase in the plasma concentration of the hormone could have lethal consequences.

Altered Responsiveness

Whether the calcium signal acts alone or in concert with cAMP, the two events of critical importance in maintaining a sustained cellular response are the association of PKC with the plasma membrane and an increased rate of calcium-ion cycling across the membrane. In the case of angiotensin II's activation of adrenal glomerulosa cells, those two events take place simultaneously. In certain circumstances, however, the events are temporally dissociated.

For example, in adrenal cells the membrane association of PKC and the increased calcium cycling can become uncoupled to produce a kind of cellular "memory." If isolated adrenal cells are perfused in the laboratory with angiotensin II for three periods of between 15 and 20 minutes, separated by intervals of similar duration, the aldosterone secretion that occurs during each successive exposure is higher than the secretion during the preceding one. Clearly, the adrenal cells "remember" their previous exposure to angiotensin II (see Figure 6.4).

That memory is transient; the longer the interval between exposures to angiotensin II, the less striking the increase in the secretion of aldosterone. The basis of this phenomenon seems to lie in the fact that the PKC associated with the membrane does not dissociate immediately when the angiotensin II signal is terminated. Additional exposure to angiotensin II not only reactivates the PKC that is still associated with the plasma membrane but also recruits additional PKC molecules to the membrane. The result is an enhanced response on reexposure to angiotensin II.

Another example of the same persistent association of PKC with the plasma membrane comes from the action of acetylcholine on beta cells. Recall that when levels of blood glucose are low, acetycholine stimulates the breakdown of PIP_2 but does not cause a significant increase in insulin secretion because it does not increase the rate of calcium cycling. What could be the role of the acetylcholine signal at that time? My colleagues the Zawalichs and I think we have an answer.

We propose that by stimulating a transient, IP_3-induced elevation of cytosolic calcium concentration and the generation of DAG, acetylcholine brings about the translocation of PKC to the plasma membrane of the beta cell. Because calcium cycling has not increased, the PKC is membrane-associated but is not activated; it is available, however, to become activated when the small, postprandial increase in glucose concentration occurs and leads to a calcium influx. Acetylcholine, then, acts to prepare beta cells to respond to a postprandial rise in blood glucose concentration with a greater release of insulin than the hormone would otherwise bring about.

Kinase Cascades

My colleagues and I extended our investigation of sustained cellular responses by looking at the contraction of smooth muscle, one of the most common tissues in the human body. Smooth muscle — which unlike skeletal muscle is not under voluntary control — is a key component of the walls of the trachea and bronchi and of the blood vessels, ureter, stomach, intestine and uterus. Susanna S.-C. Park, Yoh Takuwa, Grant G. Kelley, Hermann Haller and I have focused our studies on the smooth muscles of the trachea and carotid arteries in the cow. Acetylcholine induces a rapid and sustained increase in the contraction of tracheal muscle, and the extracellular signal histamine induces the same kind of response in the carotid artery muscle.

A model nearly identical to the one developed for angiotensin II action in adrenal cells seems to account for the action of both signals. The initiation of contraction is brought about by a transient rise in cytosolic calcium, which stimulates the calmodulin-dependent enzyme mysosin light-chain kinase. A transient increase in the extent of phosphorylation of the myosin light-chain protein follows, which initiates a rapid but transient contractile response. At the same time, the rise in cytosolic calcium (along with the DAG produced by PIP_2 breakdown) also induces PKC to associate with the membrane, as it does in adrenal cells. During the sustained phase of muscle contraction, a rise in calcium-ion concentration in the submembrane domain of the cell activates the plasma membrane-associated PKC, so that a number of proteins become and remain phosphorylated, prolonging contraction.

Two of the high-molecular-weight proteins that are phosphorylated during smooth muscle contraction, namely desmin and caldesmon, have shed light on a question presented by our new model: How does plasma membrane-associated PKC exert its effects on proteins at distant locations in the cell? Desmin and caldesmon are two such proteins, localized in domains of the cell that are remote from the site of PKC action. Given their intracellular location and their associations with complex, highly organized macromolecular structures, it is unlikely that either of these proteins shuttles between its particular domain and the membrane region where PKC operates.

Yet our own studies and those of David R. Hathaway and his colleagues at the Indiana University Medical Center show that both proteins do in fact

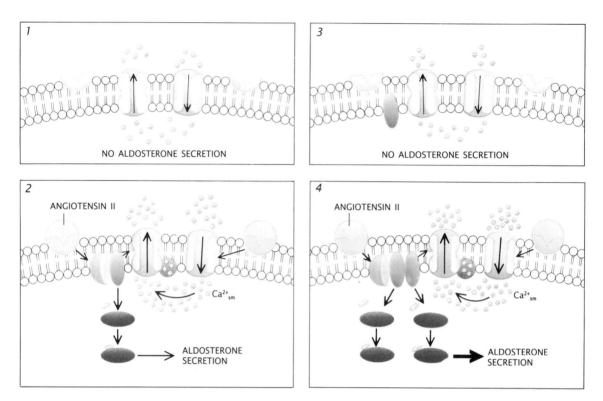

Figure 6.4 "MEMORY" effect occurs when adrenal cells are exposed to successive doses of angiotensin II in the test tube. In the basal state (*1*), calcium cycling is slow, sub-membrane calcium concentration is low and PKC is not associated with the plasma membrane. The first exposure to angiotensin II (*2*) causes PKC to move to the membrane and increases calcium cycling; these changes increase PKC activity, protein phosphorylation and the secretion of aldosterone. When angiotensin II is removed (*3*), calcium cycling decreases, but PKC remains associated with the membrane. With subsequent exposure to angiotensin II (*4*), more PKC moves to the membrane, enhancing protein phosphorylation and aldosterone secretion.

become phosphorylated during the sustained phase of tracheal or carotid artery smooth muscle contraction. In the test tube PKC will phosphorylate desmin and caldesmon directly. Close examination reveals, however, that the site of this test-tube phosphorylation is different from the site of phosphorylation during the contraction of intact muscle in response to treatment with either acetylcholine or histamine.

We derive two conclusions from these results. First, many intracellular proteins may be potential substrates for a given protein kinase, but in the cell they do not serve as such because they reside in a subcellular domain different from the one in which the kinase normally functions. Second, the phosphorylation of desmin and caldesmon in the stimulated muscle is probably achieved by a protein kinase other than PKC.

Because PKC does seem to trigger the phosphorylation, we postulate the presence of protein kinase "cascades" in which one or more of the substrates of PKC are themselves protein kinases (see Figure 6.5). Kinase phosphorylates kinase until eventually one of the kinases in the cascade phosphorylates, say, desmin or caldesmon. Workers have already seen such a cascade in the action of insulin on its target cells.

A New Calcium Messenger

In our studies of calcium-messenger function in sustained cellular responses, we have found that, contrary to the classic model of calcium as messenger, a rise and fall in the cytosolic concentration of calcium ions appears to operate as an intracellular messenger only during brief cellular responses or during the initial phases of sustained responses.

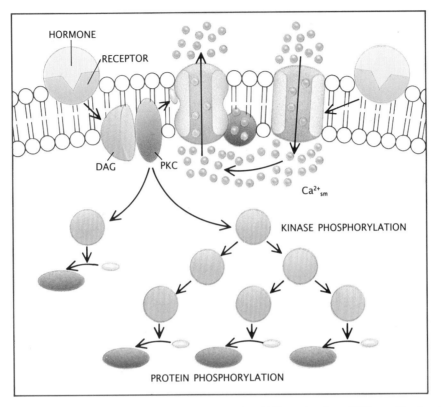

Figure 6.5 KINASE "CASCADE" may explain the ability of PKC activation to alter the phosphorylation of proteins in remote parts of the cell even though the enzyme remains at the plasma membrane. The cascade would begin with PKC activation; PKC could phosphorylate and thereby activate other kinases (*green*), which would in turn activate still other enzymes to modify the functions of many different target proteins (*red*).

During the sustained phase, a calcium signal is generated in a restricted region of the cell membrane by an increase in the rate of calcium cycling across the membrane. This submembrane calcium signal acts on calcium-sensitive, plasma membrane-associated transducers to generate other signals. By and large, it is messengers generated by the transducers— rather than calcium or the transducers themselves —that convey information from the cell surface to the cell interior.

Much remains to be learned about kinase cascades and about the separate control of calcium cycling and the plasma membrane association of PKC. Yet already a growing awareness of this new type of calcium-ion messenger has contributed to research on associative learning [see "Memory Storage and Neural Systems," by Daniel L. Alkon; SCIENTIFIC AMERICAN, July, 1989]. Because it touches on insulin secretion and blood-vessel constriction, research on the messenger should illuminate the sequence of events leading to diabetes and high blood pressure as well.

The Molecules of Visual Excitation

When a rod cell in the retina absorbs light, a cascade of reactions results in a nerve signal. That cascade has now been worked out in molecular detail. A key intermediate is a protein called transducin.

. . .

Lubert Stryer
July, 1987

This is an exciting time in the investigation of vision. Many years ago William A. H. Rushton of the University of Cambridge wrote: "Molecules respond to light as do people to music. Some absorb nothing. Others respond by the degraded vibration of foot or finger. But some there are who rise and dance and change partners." At the time Rushton wrote, his description was largely poetry: it was not known precisely which molecules are involved in the response of the retina's photoreceptor cells to light; nor was it known how those molecules interact.

In the past decade, however, experiments in the many laboratories (including my own) have revealed the molecular basic of visual excitation. The molecules that participate in the response are known, and the basic scheme of their interactions has been worked out. That detailed biochemical work has shown Rushton's description to be a prescient one. The molecules that form the basis of the response to light do indeed "rise and dance and change partners" in a remarkable cascade that lies at the root of vision.

The molecular cascade that has been worked out so carefully in the past decade has its seat in the photoreceptor cells of the retina. The photoreceptor cells are of two types, which are called rods and cones because of their characteristic shapes. Rod cells make it possible to form black-and-white images in dim light; cones mediate color vision in bright light. The human retina contains three million cones and 100 million rods. The electrical signals generated by the rods and cones are processed by other retinal cells before being transmitted to the brain by way of the optic nerve.

My interest in the molecular basis of vision was originally stimulated by certain striking properties of the rod cells. As receptors, rods have attained the ultimate in sensitivity. A rod cell can be excited by a single photon, which is the smallest possible quantity of light. The cascade of molecular reactions amplifies this minute piece of information into a signal that is useful to the nervous system. What is more, the degree of amplification varies with the background illumination: rod cells are much less sensitive in bright light than in dim light. As a result they function efficiently over a wide range of background illumination.

I was also attracted to rod cells because their exquisitely sensitive sensory system is packaged in a distinct cellular subunit that can readily be detached

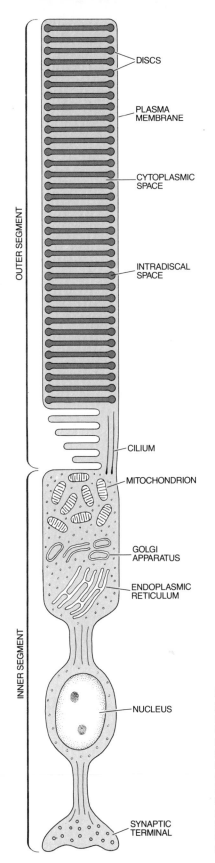

OUTER SEGMENT

DISCS

PLASMA MEMBRANE

CYTOPLASMIC SPACE

INTRADISCAL SPACE

CILIUM

MITOCHONDRION

GOLGI APPARATUS

ENDOPLASMIC RETICULUM

INNER SEGMENT

NUCLEUS

SYNAPTIC TERMINAL

Figure 7.1 ROD CELL is divided into two parts that have specialized functions. The apparatus for detecting light is in the outer segment, which holds a stack of some 2,000 disks derived from the plasma membrane. The inner segment contains organelles for making specialized molecules required in photoreception. When light strikes the disk, molecules there are modified. The signal is sent by a chain of reactions to the plasma membrane. It travels through the plasma membrane to the synaptic terminal, from which it is sent to other retinal cells.

and studied. The rod cell is a long, thin structure divided into two parts (see Figure 7.1). The outer segment contains most of the molecular apparatus for detecting light and initiating a nerve impulse. The inner segment is specialized for generating energy and renewing the molecules needed in the outer segment. In addition the inner segment includes a synaptic terminal that provides the basis for communication with other cells. If an isolated retina is gently shaken, the outer segments fall off and the machinery of excitation can be studied in a highly purified form. This feature has made the rod cell a great gift to biochemists.

The outer segment of the rod is a narrow tube

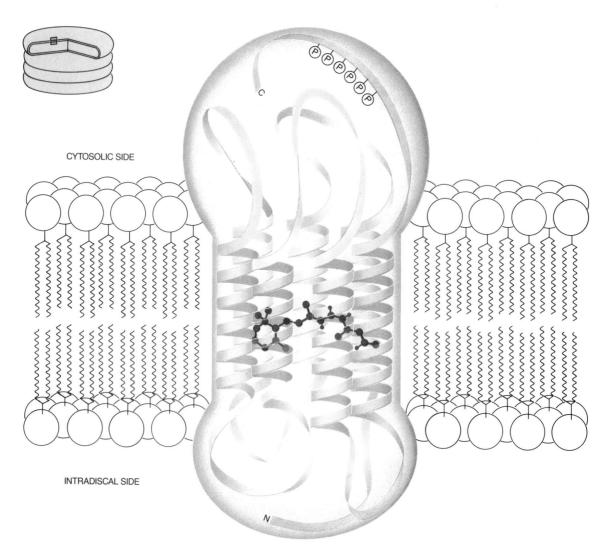

Figure 7.2 RHODOPSIN MOLECULE, embedded in the membrane of the disk, receives light and initiates the excitatory cascade. Like the plasma membrane from which it is derived, the disk membrane is a bilayer of lipid (fatty) molecules. Rhodopsin has two components: 11-*cis* retinal and opsin. Opsin is a protein that has the form of seven helical structures threaded through the membrane, connected by short linear segments. 11-*cis* retinal (**red**) lies near the center of the membrane, attached to one helix. The absorption of a photon (a quantum of light) by retinal alters retinal's shape and activates rhodopsin. This model for the structure of rhodopsin was proposed by Edward Dratz of the University of California at Santa Cruz and Paul Hargrave of Southern Illinois University.

filled with a stack of some 2,000 tiny disks. Both the tube and the disks are made up of the same type of bilayer membrane. The outer (or plasma) membrane and the disk membrane, however, have different functions in the reception of light and the generation of a nerve impulse. The stacked disks contain most of the protein molecules that absorb light and initiate the excitation response. The outer membrane serves to convert a chemical signal into an electrical one. Much of the dramatic new work on visual excitation has been devoted to tracing the process that links the molecules of the disk membrane with those of the outer membrane.

Among the most important of the molecules associated with the disk membrane is the one called rhodopsin. Rhodopsin is the photoreceptor protein of rod cells, the molecule that absorbs a photon and makes the initial response in the chain of events that underlies vision. Rhodopsin has two components, which are called 11-*cis* retinal and opsin. 11-*cis* retinal is an organic molecule derived from vitamin A. Opsin is a protein that has the capacity to act as an enzyme. The absorption of a photon by 11-*cis* retinal triggers the enzymatic activity of opsin and sets the biochemical cascade in motion.

O psin is a single polypeptide chain of 348 linked amino acids. Recently the amino acid sequence of opsin was worked out in the laboratories of Yuri A. Ovchinnikov of the M. M. Shemyakin Institute of Bioorganic Chemistry in Moscow and Paul A. Hargrave, then at Southern Illinois University. That work has provided considerable information about the three-dimensional structure of the protein, which is threaded through the disk membrane. It appears that opsin has the form of seven helixes (of the type known as alpha-helixes) arranged vertically in the membrane and connected by short nonhelical segments (see Figure 7.2). Attached to one alpha helix is a single molecule of 11-*cis* retinal, which lies near the center of the membrane, its long axis aligned with the plane of the membrane.

This arrangement leaves retinal nested at the center of a complex and highly structured protein environment. That environment (among other factors) is responsible for "tuning" retinal by influencing the spectrum of radiation it can absorb. Whereas retinal by itself in solution absorbs most intensely at a wavelength of 380 nanometers (in the ultraviolet part of the spectrum), rhodopsin does so at 500 nanometers (in the green). This shift is an excellent one from a functional point of view, because it matches the absorption spectrum of rhodopsin with the light reaching the eye.

What happens when 11-*cis* retinal absorbs a photon? The general answer is that the molecule is isomerized. Isomers are molecules having the same atoms but different shapes. Indeed, the label 11-*cis* designates a particular isomer of retinal. The backbone of retinal is a string of carbon atoms; 11-*cis* indicates that the hydrogen atoms attached to carbon atoms 11 and 12 lie on the same side of the chain. This configuration forces the chain to bend between carbons 11 and 12 (see Figure 7.3). In another isomer, called all-*trans* retinal, the hydrogens attached to carbons 11 and 12 lie opposite each other and the carbon backbone is straight.

N ow, as long ago as 1957 George Wald and Ruth Hubbard of Harvard University identified the initial molecular event in vision by showing that when 11-*cis* retinal absorbs a photon, it is converted into the all-*trans* form. The energy of light straightens out the bend in the chain of carbon atoms. In this motion rhodopsin is quite responsive: absorption of a photon leads to isomerization about half of the time. In contrast, spontaneous isomerization in the dark takes place roughly once in 1,000 years. The contrast has valuable consequences for vision. When a photon strikes the retina, the rhodopsin that is struck reports the event with high efficiency, while the millions of other rhodopsin molecules in the cell remain silent.

More than a decade after the work of Wald and Hubbard several advances revealed something about what happens at the termination of the excitatory cascade in the outer membrane. The plasma membrane is selectively permeable to ions, which carry a net electric charge. As a result there is a difference in electric potential between the inside of the rod cell and the outside. In the resting state the inside of the cell is about 40 millivolts (mV) negative with respect to the outside. In 1970 some elegant electrophysiology by Tsuneo Tomita of Keio University as well as by William A. Hagins and Shuko Yoshikami of the National Institutes of Health showed that following illumination the potential difference increases. The increase varies with the strength of the stimulus and the background illumination; the maximum potential difference is −80mV.

The increase in potential difference, which is known as a hyperpolarization, is due to a decrease in the permeability of the membrane to sodium ions

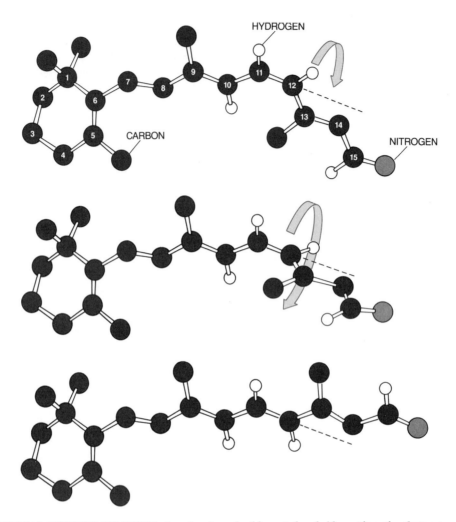

Figure 7.3 RETINAL CHANGES ITS SHAPE after absorbing a photon. In 11-*cis* retinal, the form taken in the dark (*top*), the hydrogen atoms attached to carbons 11 and 12 lie on the same side of the carbon backbone, which forces the backbone to bend. Absorption of a photon causes the chain to rotate between carbons 11 and 12. That rotation straightens out the end of the chain, yielding a different chemical form: all-*trans* retinal (*bottom*).

(which carry a positive charge). After the general nature of the hyperpolarization had been worked out, my colleague Denis A. Baylor showed that the absorption of a single photon blocks the influx of millions of sodium ions by closing hundreds of channels for sodium in the plasma membrane [see "How Photoreceptor Cells Respond to Light," by Julie L. Schnapf and Denis A. Baylor; SCIENTIFIC AMERICAN, April, 1987]. Once the sodium channels close, the light-induced hyperpolarization is passed along the outer membrane to the synaptic terminal at the other end of the cell, where the nerve impulse arises.

These fundamental results at each end of the biochemical cascade posed a stark question: What happens in between? How does the isomerization of retinal in the disk membrane lead to the closure of sodium channels in the outer membrane? The plasma membrane of the rod cell is physically distinct from the disk membranes. Hence the signal must be carried from the disks to the outer membrane by a transmitter. Since the absorption of a single photon can lead to the closing of hundreds of sodium channels, many transmitters must be formed per photon absorbed.

What is the transmitter that carries the excitation

signal? In 1973 Hagins and Yoshikami proposed that calcium ions, sequestered in the disk in the dark, are released on illumination and diffuse to the plasma membrane to close sodium channels. This attractive hypothesis generated much interest and many experiments. Recent work has shown, however, that although calcium ion has a significant role in vision, it is not the excitatory transmitter. Instead the transmitter is called 3′,5′ cyclic guanosine monophosphate, or cyclic GMP (see Figure 7.4).

The capacity of cyclic GMP to act as a transmitter is closely related to its chemical structure. GMP is a nucleotide of the type that forms the subunits of RNA. Like other nucleotides, it has two components: a base and a five-carbon sugar unit. In the case of GMP the base is guanine; nucleotides containing guanine are known as guanyl nucleotides. The word cyclic indicates that the carbons designated 3′ and 5′ in the sugar molecule are joined by a phosphate group. The link that joins the two carbons—known as a phosphodiester bond—forms a ring. When the ring is intact, cyclic GMP is capable of keeping the membrane sodium channels open. When it is cleaved by an enzyme called a phosphodiesterase, the sodium channels close spontaneously (see Figure 7.5).

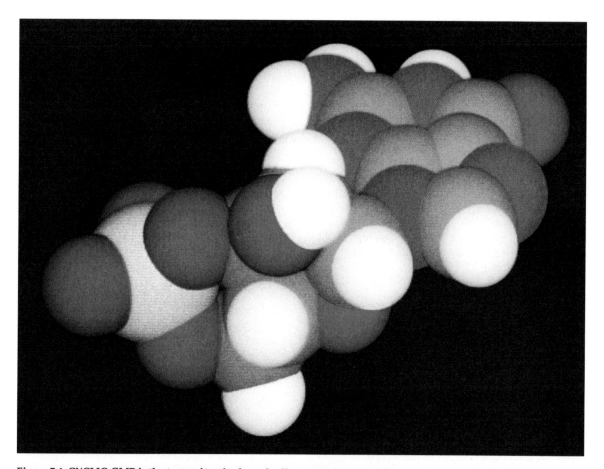

Figure 7.4 CYCLIC GMP is the transmitter in the rod cells of the retina that is directly responsible for generating a nerve impulse. Cyclic GMP includes atoms of nitrogen (*dark blue*), carbon (*light blue*), hydrogen (*white*), oxygen (*red*) and phosphorus (*yellow*). The phosphorus atom forms part of the ring structure for which the molecule is called cyclic. When the ring is intact, cyclic GMP holds open channels for sodium ions in the outer membrane of the rod cell. When the ring is enzymatically cleaved, the sodium channels close and the electrical properties of the membrane change, giving rise to a nerve impulse.

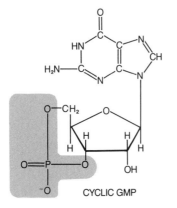
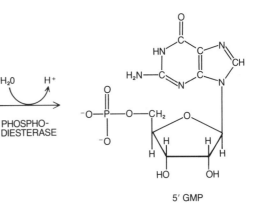

CYCLIC GMP

5' GMP

Figure 7.5 RING OF CYCLIC GMP is opened by an enzyme called a phosphodiesterase. The ring (*color*) includes a phosphorus atom in a link called a phosphodiester bond. The enzyme inserts a water molecule into the bond, cleaving it to yield a molecule called 5' GMP.

S everal steps intervene between the excitation of rhodopsin and the enzymatic cleavage of cyclic GMP. When retinal absorbs a photon and opsin is activated, rhodopsin in turn activates an enzyme called transducin. Transducin, whose action was first elucidated in my laboratory, is a key intermediate in the excitatory cascade. Transducin in turn activates a specific phosphodiesterase. The phosphodiesterase then opens the ring of cyclic GMP (see Figure 7.5) by inserting a water molecule into it (a process known as hydrolysis). Although this pathway is not difficult to describe in outline, unraveling it and understanding its physiological significance required a wide variety of experiments in many laboratories.

In 1971 Mark W. Bitensky and William H. Miller of the Yale University School of Medicine found that light markedly reduces the level of a cyclic nucleotide in rod outer segments. Subsequent studies showed that the reduction was due to the hydrolysis of cyclic GMP by a phosphodiesterase specific to that nucleotide. At that time the calcium hypothesis was still quite strong, however, and it was by no means clear that cyclic GMP had much direct influence on the excitatory response. Then in the late 1970's two seminal findings were made; those were the ones that initially aroused my interest in cyclic GMP as a transmitter candidate.

At a research conference in the summer of 1978 Paul A. Liebman of the University of Pennsylvania reported his finding that a single photon could trigger the activation of hundreds of phosphodiesterase molecules per second in preparations of rod outer segments. Earlier work, carried out in the presence of the nucleotide adenosine triphosphate (ATP), had shown much less amplification. Liebman observed that considerably more amplification could be obtained with guanosine triphosphate (GTP).

Guanosine triphosphate is a nucleotide closely related to the noncyclic form of GMP. Instead of having a single phosphate group attached to its 5' carbon, however, it has a chain of three phosphates bound to each other by phosphodiester linkages. The energy stored in those bonds provides the basis for many cellular functions. For example, the splitting off of one phosphate group converts GTP into guanosine diphosphate (GDP) and liberates considerable energy. In this way the cell obtains energy needed to drive chemical reactions that are otherwise energetically unfavorable. Liebman's key finding was that this process seemed to be at work in the activation of the phosphodiesterase, where GTP is an essential cofactor.

On the way home from the research conference where Liebman had presented his report, I visited Yale and had the good fortune to stay overnight at Miller's home. After an excellent dinner he showed me some intriguing experimental records. Miller and his co-worker Grant Nicol had succeeded in injecting cyclic GMP into the outer segment of intact rod cells. The results were striking. Cyclic GMP quickly reduced the potential difference across the plasma membrane. Not only that, the injected nucleotide sharply increased the delay between the arrival of a light pulse and the hyperpolarization of the membrane. The simplest interpretation was that cyclic GMP opened sodium channels, which then remained open until cyclic GMP was degraded by the light-activated phosphodiesterase.

That hypothesis was an alluring one, but direct proof was lacking. On returning to my laboratory I discussed the new findings with Bernard K.-K. Fung, a postdoctoral fellow. Fung shared my enthusiasm for cyclic GMP as a transmitter. Together we decided to explore the molecular mechanism by which the phosphodiesterase that cleaves cyclic GMP is activated. Liebman's discovery that GTP was necessary for activation was significant in our thinking, because it suggested that a protein capable of binding GTP might be a significant intermediate in activation. We began to look carefully at what happens to GTP in rod cells.

Our first experiment was designed to detect the binding of GTP and its chemical derivatives to rod outer segments. Radioactively labeled GTP was incubated with rod outer-segment fragments. After several hours the preparation was washed over a filter that retained membrane fragments or molecules as large as proteins but allowed small molecules such as GTP (and its relatives) to pass. We found that a substantial amount of the radioactivity was bound to the membrane. Further work showed, however, that the molecule bound to the membrane was not GTP but GDP.

Those results strongly suggested that there was a protein in the rod-cell membranes that is capable of binding GTP and then converting it into GDP by cleaving off one of its phosphate groups. It seemed increasingly likely that such a protein was a key intermediate and also that the conversion of GTP into GDP might be driving the activation process. While this work was in progress Walter Godchaux III and William F. Zimmerman of Amherst College reported finding just such a protein in rod-cell membranes. It was not clear, however, how their finding fitted with the other pieces of the puzzle that were then emerging.

One of the striking aspects of the activity Fung and I observed was that not only was there something in the membrane capable of binding guanyl nucleotides but also when the membranes were illuminated, GDP was released. Moreover, the release of GDP from the membrane was strongly enhanced by the presence of GTP in the surrounding solution. Now a hypothesis began to emerge to simplify this welter of data. It seemed one part of the activation process involved an exchange of GTP for GDP in the membrane. That was why the release of GDP was so markedly enhanced by GTP: the release of a GDP molecule depended on the substitution of a GTP for it. Later perhaps GTP might be converted into GDP.

More and more, the exchange of GTP for GDP seemed to be near the heart of the activation process. For one thing, that exchange is highly amplified. We measured the effect of light on the uptake of GTP by the membrane and found that the photoexcitation of one rhodopsin molecule led to the binding of about 500 molecules of a GTP analogue (chosen because it resists being reduced to GDP and therefore enabled us to isolate the exchange step). The discovery of this amplification was exciting, because it pointed toward an explanation of the overall amplification characterizing the excitatory cascade.

This central result led us to propose that there is a protein intermediate in the excitation cascade that can exist in two states. In one state the protein binds GDP; in the other it binds GTP. The substitution of GTP for GDP is the signal for the protein's activation. That exchange is triggered by rhodopsin; in turn it serves to activate a specific phosphodiesterase. The phosphodiesterase then closes the sodium channels in the plasma membrane by cleaving GMP. Building on the work of Hermann Kühn of the Institute for Neurobiology at the University of Jülich, we soon isolated the postulated protein. It was given the name transducin in recognition of the fact that it mediates the conversion of light into an electrical impulse, a process known as a transduction. Subsequent work showed that transducin consists of three protein subunits designated alpha, beta and gamma.

Having the purified transducin in hand enabled us to test our hypothesis about the overall flow of information in the activation cascade. We had suggested that the signal moved from activated rhodopsin to transducin (in its GTP form) to the phosphodiesterase. If that picture was correct, two things followed. First, it could be inferred that transducin could be converted into its GTP form in the absence of the phosphodiesterase. It also followed that the phosphodiesterase could be activated in the absence of photoexcited rhodopsin.

My colleagues and I set out to test both proposi-

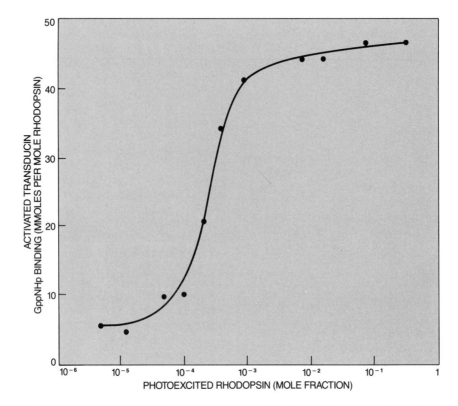

Figure 7.6 ACTIVATION OF TRANSDUCIN is accompanied by the binding of guanosine triphosphate (GTP). The horizontal axis shows the concentration of photoexcited rhodopsin (increasing to the right). The vertical axis shows the binding of GppNHp (a GTP analogue) in fragments of artificial membrane containing only rhodopsin and trans- ducin. The slope of the curve indicates that each rhodopsin molecule activates 71 molecules of transducin in this system. (In intact rod cells 500 molecules of transducin are activated by each rhodopsin.) Activated transducin triggers the phosphodiesterase that cleaves cyclic GMP.

tions. To test the first we assembled a synthetic-membrane system from which the phosphodiesterase was absent. Purified transducin in the GDP form was introduced into the artificial membranes. When rhodopsin was activated by light, we found that each molecule of rhodopsin catalyzed the uptake of 71 molecules of a GTP analogue to the membrane (see Figure 7.6). Thus it appeared that each molecule of rhodopsin was activating transducin by catalyzing the exchange of GTP for GDP in many molecules of transducin. This was the first time an amplified effect of rhodopsin required only one other protein—transducin—had been found.

We then wanted to test the second proposition: that transducin in the GTP form can activate the phosphodiesterase in the absence of any photoexcited rhodopsin. The first step was to purify the

active form of the enzyme: the transducin-GTP complex. When we did so, we got a surprise. We knew that in the inactive, GDP form transducin was complete; all three subunits were joined. In the active, GTP form, however, transducin came apart. The alpha subunit floated free of the joined beta and gamma units. GTP was bound to the alpha subunit.

We were now in a position to ask an even more precise question than the one we had intended to ask. We could now inquire whether it is the alpha subunit (with its attached GTP) or the beta-gamma subunit that stimulates the action of the phosphodiesterase. We obtained a decisive answer. The alpha subunit (with GTP) activates the phosphodiesterase; the joined beta-gamma unit has no effect (see Figure 7.7). Moreover, the alpha unit was effec-

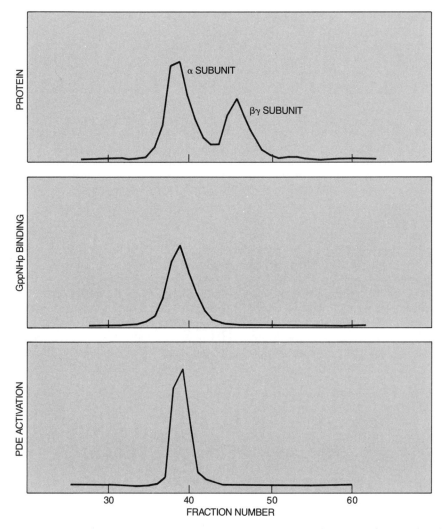

Figure 7.7 ALPHA SUBUNIT OF TRANSDUCIN is the information carrier in the activation of the phosphodiesterase. Transducin has three subunits, which are designated alpha, beta and gamma. The three graphs show the result of passing a solution containing transducin through a filtration column. The horizontal axis indicates the array of equal portions of fluid that were collected at the bottom of the column. The top panel shows that the protein in the solution has two components of differing size. The small component is the alpha subunit; the large one is the joined beta-gamma unit. The middle panel indicates that the GppNHp-binding activity is associated with the alpha subunit. The bottom panel indicates that phosphodiesterase activation is also associated with the alpha subunit.

tive in the absence of rhodopsin, validating the idea that transducin can activate the phosphodiesterase without rhodopsin and confirming our overall picture of the information flow.

What is the mechanism by which transducin activates the specific phosphodiesterase?

Bitensky's group found that in the dark the phosphodiesterase has a low activity because it is subject to an inhibitory constraint. When a small amount of trypsin (a protein-digesting substance) was added, the constraint was removed, activating the enzyme. It was known that the phosphodiesterase consists of three polypeptide chains; as in the case of transducin, they are designated alpha, beta and gamma. James B. Hurley, a postdoctoral fellow in my laboratory, found that trypsin degrades the gamma subunit but not the alpha or beta units. Taken together, Bitensky's and Hurley's results made it seem likely that the inhibitor of the phosphodiesterase is the gamma subunit.

That notion was confirmed by results from my laboratory and others. My co-workers and I purified the gamma subunit, added it to the active alpha-beta complex and found that gamma eliminated more than 99 percent of the catalytic activity. Further confirmation came from the fact that the rate of destruction of the gamma unit by trypsin closely resembled the rate of activation of the phosphodiesterase in the excitatory cascade. Finally, Marc Chabre and his colleagues at the Center for Nuclear Studies at Grenoble found that transducin in the GTP form can bind to the gamma subunit of the phosphodiesterase and form a complex with it.

The picture that emerges from these results is that after illumination the alpha subunit of transducin with its attached GTP binds to the phosphodiesterase and carries away the inhibitory gamma subunit. The departure of gamma unleashes the catalytic activity of the phosphodiesterase. That activity is powerful: each activated enzyme molecule can hydrolyze 4,200 molecules of cyclic GMP per second.

With these insights much of the activation cascade became clear. The first step is the activation of transducin by photoexcited rhodopsin. In that interaction rhodopsin caused the attached GDP to break loose from transducin. A molecule of GTP takes its place, and the alpha subunit dissociates from the rest of the protein, taking the GTP with it. This process takes only about a millisecond, as Chabre showed with T. Minh Vuong, a graduate student, during a sabbatical spent in my laboratory. The production of hundreds of active alpha-transducin-GTP complexes by a single rhodopsin is the first stage of amplification in vision.

Alpha-transducin with its GTP then triggers the activity of the phosphodiesterase. At this stage there is no amplification: each alpha-transducin unit binds to and activates a single phosphodiesterase. The transducin-phosphodiesterase pair act as a single unit to provide the second amplified stage. The transducin remains associated with the phosphodiesterase as it does its work of cleaving cyclic GMP. As I described, each activated enzyme molecule can cleave several thousand cyclic GMP's. This amplification, along with the activation of many transdu-

cins by each rhodopsin, accounts for the remarkable magnification of a single photon into a palpable nerve impulse.

Yet if the organism is to be able to see more than once, this cycle must also be turned off. Transducin has a key role in deactivation as well as in activation. The alpha subunit has a built-in chemical timer that terminates the activated state by converting the bound GTP into GDP. The mechanism of the timer is not fully understood. It is known, however, that the hydrolysis of GTP into GDP in the deactivation phase has an important role in driving the entire cycle. The activation reactions are energetically favorable. Some of the deactivation reactions, on the other hand, are not; without the conversion of GTP into GDP the system could not be reset for future firing.

As GTP is cleaved to form GDP, the alpha unit of transducin releases the inhibitory gamma unit of the phosphodiesterase. The gamma unit returns to the phosphodiesterase, binds to it and restores it to the quiescent state. Transducin is then restored to its preactivation form by the rejoining of the alpha subunit and the beta-gamma unit. Rhodopsin is deactivated by an enzyme that recognizes its specific structure. That enzyme, called a kinase, attaches multiple phosphate groups to amino acids at one end of the opsin chain. As Kühn has shown, rhodopsin then forms a complex with a protein called arrestin, which blocks the binding of transducin and puts the system back in the dark state.

Much of the working out of the visual cascade that was done in the late 1970's and early 1980's proceeded on the assumption that cyclic GMP somehow opens the sodium channels in the outer membrane and that hydrolysis of cyclic GMP leads to their being closed. Little was known, however, about the mechanism by which this might take place. Was cyclic GMP acting directly on the channels, or did it act through intermediaries? A decisive answer was obtained in 1985 by Evgeniy E. Fesenko and his co-workers at the Institute of Biological Physics in Moscow.

Fesenko employed a micropipette to pull off a small patch of the rod cell's plasma membrane. The patch adhered tightly to the end of the pipette, with the side that would normally be inside the cell facing out. This face of the membrane was then exposed to various solutions in order to test their effect on the sodium conductance. The results were unambiguous: the channels were directly opened by cyclic GMP but were not affected by other substances, including calcium ion.

Fesenko's incisive experiments gave the final blow to the notion that calcium ion might be the excitatory transmitter and established the last link in the excitatory cascade. The outline of the excitation pathway has been delineated (see Figure 7.8); as we had hypothesized, the overall information flow is from rhodopsin to transducin to the phosphodiesterase and then to cyclic GMP.

Although working out the pathway and mechanism of the excitatory cascade was very gratifying, several important questions have not yet been answered. One of them concerns how the response of

the cascade is modulated. As I mentioned above, rod cells are much less sensitive in bright light than they are in the dark. The background lighting must somehow affect the degree of "gain" in the system: the total amount of amplification provided by the two amplified steps (from rhodopsin to transducin and from the phosphodiesterase by cyclic GMP). Much evidence suggests that calcium ion has a role in the process, but the details are not known.

Other questions involve the structure of the sodium channels and the mechanism that prevents the ultimate depletion of cyclic GMP in the cell. Liebman's group and that of Benjamin Kaupp

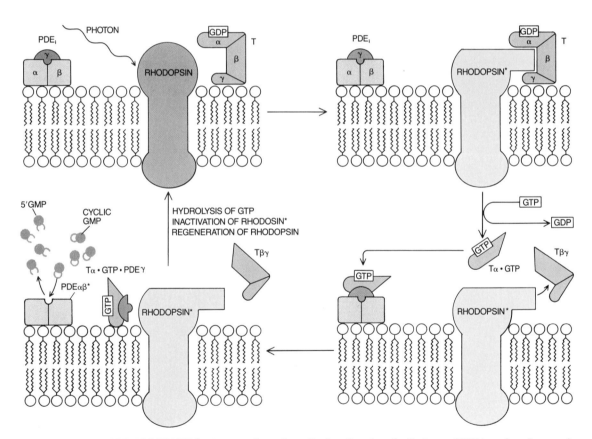

Figure 7.8 EXCITATORY CASCADE forms part of a cycle mediated by transducin. The cycle begins with the absorption of a photon by rhodopsin (*upper left*). Activated rhodopsin (*rhodopsin**) then interacts with transducin (T), as shown at the upper right. That interaction causes GTP to replace GDP in the alpha subunit of transducin and causes the alpha unit to split off from the beta-gamma part of the enzyme (*lower right*). Transducin relieves the inhibition of

the inactive phosphodiesterase (PDE$_1$), perhaps by carrying away its gamma unit. The activated phosphodiesterase begins cleaving many molecules of cyclic GMP (*lower left*). Before long a built-in timer in the alpha subunit of transducin cleaves GTP to GDP. The alpha subunit rejoins the beta-gamma unit; the phosphodiesterase is also reassembled. At the same time rhodopsin is inactivated and then regenerated in its preactivation form.

of the Neurobiology Institute at the University of Osnabrück in West Germany have made important contributions to the first question by purifying cyclic-GMP-gated channels and reconstituting their function in model membranes. A key molecule in the answer to the second question must be guanylate cyclase, the enzyme that synthesizes cyclic GMP. Clearly there must be a feedback loop that restores cyclic GMP to its preillumination level; if there were not, the cell could fire only a few times before permanently exhausting its own capacity. The nature of the loop, however, is not known.

In addition to clearing up the remaining questions about the visual cascade in rod cells, current work is extending the recent findings beyond those cells. Cone cells have a transduction system resembling that of rods, as has been shown by Baylor as well as by King-Wai Yau of Johns Hopkins University. It has long been known that the cones contain three visual pigments (analogous to rhodopsin) that respond to red, green and blue light. All three pigments contain 11-*cis* retinal. Furthermore, molecular-genetic studies recently carried out by my colleagues Jeremy Nathans and David S. Hogness have shown that the three cone pigments have the same fundamental architecture as rhodopsin. Transducin, the phosphodiesterase and the cyclic-GMP-controlled channel in the cones are like their counterparts in the rods. Hence it may not be long before the transduction cycle there is understood in the same molecular detail as it is in the rod cells.

The cyclic-GMP cascade has a significance that extends even beyond vision. The excitatory cascade in the rod cells has a notable resemblance to the pathway by which certain hormones yield their effects. For example, the hormone epinephrine (adrenaline) acts by triggering the activation of an enzyme known as adenylate cyclase. Adenylate cyclase catalyzes the formation of cyclic adenosine monophosphate (cyclic AMP). Cyclic AMP is an intracellular messenger that mediates the action of many hormones (see Chapter 4, "The Molecular Basis of Communication within the Cell," by Michael J. Berridge).

Alfred G. Gilman of the University of Texas Health Science Center at Dallas has worked out many of the facets of the regulation of adenylate cyclase, which turns out to have striking parallels to the excitatory cascade in the rod cell. Just as the excitatory cascade begins when rhodopsin absorbs a photon, the cascade of hormone action is initiated when a receptor on the cell's surface binds a specific hormone (see Figure 7.9). The receptor-hormone complex then interacts with a G protein that resembles transducin. The same exchange of bound molecules that activates transducin—GTP for GDP—activates the G protein when the G protein interacts with the receptor-hormone unit.

The parallelism does not end there. Like transducin, the stimulatory G protein has three subunits. It is the alpha subunit that activates adenylate cyclase by relieving an inhibitory constraint. Again as in the case of transducin, the G protein's stimulatory action is turned off by a built-in timer that converts GTP into GDP.

The similarities between transducin and the G proteins (several have been identified) apply to their structure as well as to their activity. Transducin and the G proteins are members of the same overall family of signal-coupling proteins. All members of that family identified so far have nearly the same beta subunit. In addition, their alpha subunits have the same function, a similarity that has now been shown at the molecular level. The stretches of DNA encoding the alpha subunits of transducin and three G proteins have recently been sequenced in several laboratories. The information from the DNA indicates that about half of the amino acid sequences in the four proteins are identical or nearly so.

When the genetic information is examined on an overall level, it is found that the proteins include some regions that have been conserved quite stringently through evolution as well as some that have diverged widely. Each protein contains three binding sites: one for guanyl nucleotides, one for the activated receptor (rhodopsin or a hormone-receptor complex) and one for the effector protein (phosphodiesterase or adenylate cyclase). As might well be expected—given its crucial function in the activation cascade—the binding site for GTP or GDP is the most highly conserved among the various proteins.

Indeed, the GTP-binding regions of these proteins resemble a region of a functionally very different protein called elongation factor *Tu*. *Tu* has a crucial role in the synthesis of proteins in certain bacteria; it delivers transfer RNA's (the "hooks" that carry specific amino acids and allow them to be added to a growing amino acid chain) to the ribosome, which is the organelle where proteins are synthesized. In its work *Tu* undergoes a cycle akin to the transducin cycle in visual activation. The cycle is powered by the cleaving of a molecule of GTP, and there is a site on *Tu* where the GTP binds.

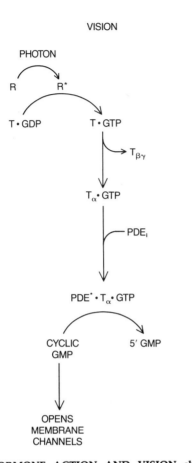

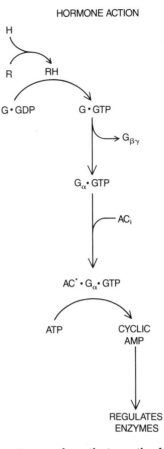

Figure 7.9 HORMONE ACTION AND VISION show striking parallels. The sequence at the left shows the excitatory cascade of vision; the one at the right shows the events by which a variety of hormones have their effects. The binding of a hormone (H) and its receptor (R) to form a hormone-receptor complex resembles the excitation of rhodopsin. As rhodopsin activates transducin, the hormone-receptor complex activates a stimulatory G protein. The G protein activates adenylate cyclase (AC), which is analogous to the phosphodiesterase. Adenylate cyclase catalyzes the conversion of adenosine triphosphate (ATP) to cyclic adenosine monophosphate (cyclic AMP), which regulates many enzymes.

That site is quite close in amino acid sequence to the guanyl-nucleotide binding sites of transducin and the various G proteins.

Now protein synthesis is one of the most fundamental metabolic activities of any cell. Therefore it seems likely that elongation factor Tu, which takes part in that work, originated earlier in evolution than the G proteins or their relative, transducin. Indeed, Tu could be the ancestor of both transducin and the G proteins. The controlled uptake and release of proteins coupled to GTP-GDP exchange (and subsequent cleavage) was undoubtedly perfected early in evolution: elongation factor Tu may represent an early version of the cycle.

One of the many fascinating things about evolution is that mechanisms evolved for a particular function may later be modified and applied to different functions. That, I think, is what happened to the mechanism of Tu. After evolving to mediate protein synthesis, it was retained for billions of years and ultimately put to work in the transduction of hormonal and sensory stimuli. In the past few years one of those functions—the transducin cycle—has been worked out in great detail. The result has been extremely satisfying. For the first time we now understand at the molecular level one of nature's most precise sensory events: visual excitation.

How Receptors Bring Proteins and Particles into Cells

Receptor-mediated endocytosis is a process whereby cells can take up specific large molecules. In most cases a receptor, having delivered its ligand, is recycled to the plasma membrane to bind more ligand.

. . .

Alice Dautry-Varsat and Harvey F. Lodish
May, 1984

The cells of a multicellular organism are surrounded by an aqueous medium, derived from the blood, that is like a very rich soup. It is an unusual soup in that it has many thousands of ingredients, most of them present at exceedingly low concentrations. Some of the ingredients are cellular building materials such as amino acids and nutrients such as vitamins, each of which is needed by a given cell in particular quantities and at particular times. Some ingredients are hormones delivering specific intercellular signals. Some are waste products or even toxic substances that particular cells are equipped to break down. Each cell must extract from the extracellular medium the substances it needs to internalize, rejecting the rest.

The sieve is the plasma membrane, which bounds the cell and controls its traffic with the medium and thus with every other cell of the organism. Like all biological membranes, the plasma membrane is mainly a double layer of phospholipid molecules in which many kinds of protein molecules are embedded. The proteins have a wide range of functions, one of which is to facilitate the selective uptake of specific water-soluble substances through the otherwise impermeable lipid bilayer. Ions (charged atoms) and small water-soluble molecules such as amino acids (the constituents of proteins) and sugars simply flow through or are pumped through specialized channels in the membrane; the channels are composed of proteins called permeases. Large molecules and particulate matter are brought into the cell by a quite different process, called endocytosis: a patch of the plasma membrane surrounds the material to be taken in, which is thus brought into the cell enclosed in a vesicle derived from the plasma membrane. There are three kinds of endocytosis (see Figure 8.1).

In phagocytosis the binding of a very large particle or molecular complex to the surface of the cell triggers an expansion of the membrane around the object, which is incorporated into the cell in a baglike vesicle, an invaginated patch of the membrane, that can be several micrometers in diameter. Phagocytosis is the process whereby protozoans ingest bacteria and other food particles; in higher animals the immune-system cells called macrophages engulf intruders such as bacteria by phagocytosis.

Pinocytosis is a different process that results in

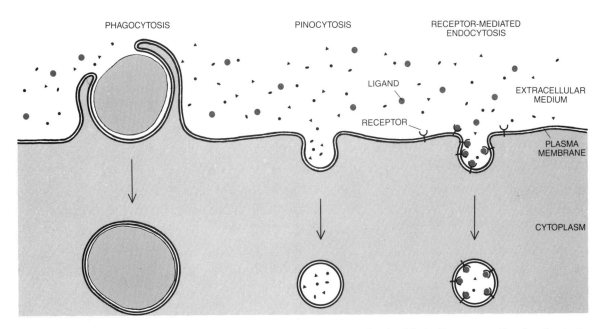

Figure 8.1 THREE KINDS OF ENDOCYTOSIS are diagrammed schematically. In phagocytosis (*left*) the cell's plasma membrane binds to a large particle such as a bacterium and expands to wrap around it and engulf it within the cell. In pinocytosis (*center*) a droplet of the liquid extracellular medium is surrounded by a patch of the membrane, which folds inward and "buds off" to form a membrane-bounded vesicle enclosing the droplet and any small molecules dissolved in it. Receptor-mediated endocytosis (*right*) is a mechanism for the selective uptake of large molecules or particles. A ligand binds to its specific receptor on the plasma membrane, triggering the internalization of the receptor-ligand complex in an invagination of the plasma membrane. The vesicle thus formed buds off inward, carrying the ligand into the cell.

the nonspecific uptake of extracellular fluid. A minute droplet of liquid is surrounded by a bit of invaginated plasma membrane and is internalized in a vesicle only about .1 micrometer in diameter, bringing with it whatever ions or small molecules happen to be in the droplet.

Receptor-mediated endocytosis, in contrast, is exquisitely specific. The receptors are membrane proteins, each of which has a binding site that fits a particular ligand: a protein or a small particle. The receptor in effect plucks one ingredient from the extracellular soup—even if it is present in a very low concentration and with a vast excess of unrelated molecules—and holds it fast. The binding triggers an invagination of the membrane to form a membrane-bounded vesicle enclosing the ligand and so to bring the ligand into the cell. In the past few years much has been learned about the mechanism of receptor-mediated endocytosis, and in particular about the remarkable events whereby the internalized receptor-ligand complex is dissociated,

the ligand is dispatched to its intracellular destination and the receptor is recycled to the surface of the cell to bind more of the ligand.

A wide variety of ligands are brought into cells by receptor-mediated endocytosis, and once in a cell they are disposed of differently. The first example of protein uptake by this method was reported in 1964 by Thomas F. Roth and Keith R. Porter, who were then at Harvard University. It explained how insect and bird eggs acquire the large amount of protein stored in the yolk body to nourish the embryo: the proteins are imported into the developing oocyte after having been synthesized elsewhere (see Figure 8.2). Roth and Porter showed that in the mosquito the precursor protein vitellogenin is synthesized in the female's liver, secreted into the blood and carried to the ovary, where it binds to receptors in some 300,000 tiny pits on the surface of the oocyte. The pits are internalized as vesicles, which fuse to form the yolk body. The

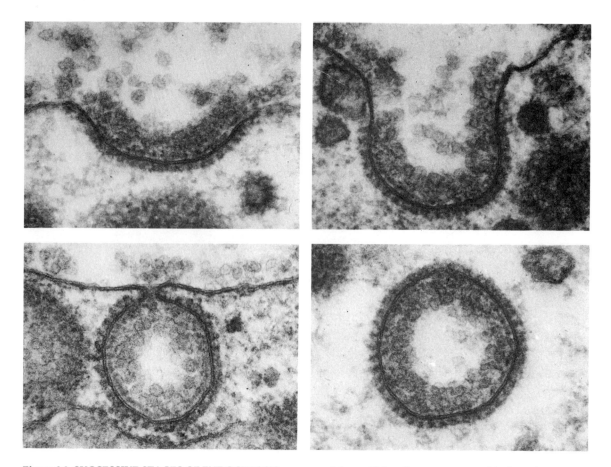

Figure 8.2 SUCCESSIVE STAGES OF ENDOCYTOSIS are seen in a series of micrographs of hen oocytes, the precursors of eggs. The shallow indentation in the oocyte's plasma membrane (*top left*) is a coated pit. The thick coating on the cytoplasmic (inner) side of the pit is a lattice of the protein clathrin. Large particles lining the membrane on the extracellular side of the pit are thought to be lipo- protein particles; they are presumably bound to specific receptors embedded in the oocyte's membrane. At later stages the pit deepens (*top right*) and forms a nascent coated vesicle (*bottom left*), and the vesicle is pinched off from the surface membrane, internalizing the lipoprotein ligand (*bottom right*).

vitellogenin is cleaved by protein-digesting enzymes (proteases) to yield two essential yolk proteins, lipovitellin and phosvitin.

In mammals maternal immunity is transferred to the developing fetus by receptor-mediated endocytosis. Antibodies from the mother's blood bind to fetal cells lining the yolk sac, which have surface receptors that recognize a region common to all circulating antibodies of the immunoglobulin-G type. The antibodies are transported into the cells and are secreted into the fetal circulation.

Cholesterol is a major lipid component of the plasma membrane of all mammalian cells. Richard G. W. Anderson, Michael S. Brown and Joseph L. Goldstein of the University of Texas Health Science Center at Dallas have shown in elegant detail how the cholesterol is made available to the cell. Cholesterol is synthesized and stored in the liver and there packaged into a large, spherical low-density lipoprotein (LDL) particle, which serves as its transporter. Each LDL particle has a core of some 1,500 cholesterol molecules chemically bound to fatty-acid chains. The core is enveloped in a single-layer phospholipid membrane and more cholesterol; a binding protein called *apo-b* is embedded in the membrane and the particle is secreted into the circulation.

LDL particles bind to specific cell-surface receptors that recognize the *apo-b*, are internalized in membrane-bounded vesicles and then transported, in a series of progressively larger vesicles, to the lyosome: a cell organelle equipped with a number of hydrolases, or digestive enzymes. The enzymes break down the LDL particles, degrading the binding protein, splitting the fatty acid from the cholesterol molecules and releasing the cholesterol and the fatty acids for incorporation into the cell membrane. Brown and Goldstein and their colleagues have shown that a lack of functional LDL receptors is responsible for familial hypercholesteremia, a genetic disease characterized by extremely high blood-cholesterol levels and consequent early atherosclerosis (see Chapter 13).

Iron is an essential constituent of all cells. It is brought to cells by a carrier protein called transferrin. The carrier binds ferric ions (Fe^{+++}) in the intestine (where iron is absorbed from foods) and in the liver (where iron is stored). A loaded ferrotransferrin molecule, carrying two iron ions, binds to a specific receptor on the surface of a cell, and the receptor-transferrin complex is internalized by endocytosis, making the iron available to the cell (see Figure 8.3).

One of the major roles of receptor-mediated endocytosis is to internalize hormones and other proteins that deliver specific signals to certain cells. Insulin molecules, for example, bind to specific receptors on a target cell. The binding triggers several metabolic processes, including an increase in the uptake of glucose. It is thought the hormone response is terminated when the insulin-receptor complexes are internalized and the insulin is degraded in lysosomes. Potentially harmful materials are also disposed of in this way. For example, there are abnormal glycoproteins (proteins with attached sugar chains) that have the sugar galactose at the terminus of the chain instead of the usual sugar, sialic acid. Liver cells have receptors that recognize and bind galactose-terminal glycoproteins, which are then taken into the cell to be broken down in lysosomes.

The receptors that mediate endocytosis are proteins that span the thickness of the plasma membrane. Receptor proteins are amphipathic: they have two hydrophilic regions, which extend into the aqueous medium outside the cell and within the cytoplasm, and a central hydrophobic region that binds tightly to the fatty acids forming the core of the membrane. Not much is known about the structure of these receptors, but the receptor for transferrin is known to be a glycoprotein with a molecular weight of 180,000. Ian S. Trowbridge of the Salk Institute for Biological Studies, Robert Allen of the University of Colorado and Howard H. Sussman of Stanford University have shown that the transferrin receptor is made up of two identical polypeptide chains, each one about 800 amino acids long, that are linked by a single disulfide bond. In addition to at least three carbohydrate chains each polypeptide carries a fatty acid, palmitate, that may help to anchor the receptor in the membrane.

Each receptor binds two molecules of transferrin, presumably one molecule per polypeptide chain. Under physiological conditions the binding is followed quickly by internalization. The two steps can be separated for experimental purposes because binding does not require energy, whereas internalization does require it and can therefore be blocked by low temperature or by inhibitors of cellular energy production: under these conditions binding takes place normally and can be quantified. Like most receptors, the one for transferrin has a very high affinity for its ligand (see Figure 8.4). Its dissociation constant (the concentration of ligand at which half of the receptors are occupied) is 5 nanomolar (5×10^{-9}M), or some 350 micrograms per liter. This means that transferrin is bound to its receptor even when the transferrin concentration is only about one hundred-thousandth of the total concentration of proteins in the blood.

Ligand-receptor complexes cluster at particular sites on the plasma membrane, namely in what are called coated pits. The pits, first observed by Roth and Porter, turn out to be present on almost all animal cells; typically they account for some 2 percent of the cell surface. The coat for which the pits are named, a thick proteinaceous layer on the inner side of the plasma membrane under each pit, is mainly clathrin, a fibrous protein that was identified by Barbara M. F. Pearse of the Medical Research Council Laboratory of Molecular Biology in England.

It is the coated-pit region of the membrane that invaginates to form a vesicle. Indeed, the process is continual. Coated pits keep folding inward to form vesicles and are continually regenerated on the cell surface. The vesicles are themselves coated, but

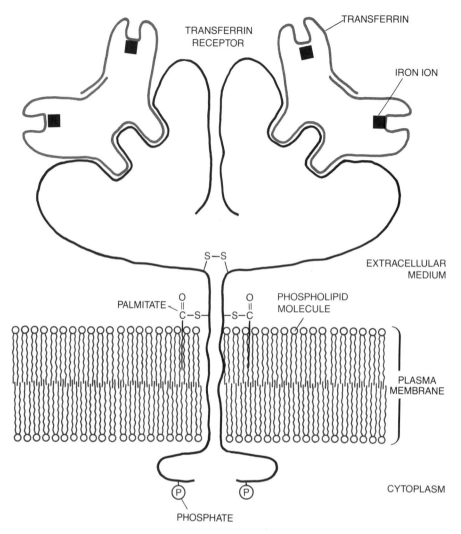

Figure 8.3 RECEPTOR FOR TRANSFERRIN, an iron-carrying protein, is embedded in the plasma membrane of a cell. The membrane is mainly a bilayer of phospholipid molecules whose fatty-acid tails make the membrane impermeable to water-soluble molecules. The actual structure of the transferrin receptor is not known; its shape in this drawing reflects what is known. The receptor has two identical polypeptide chains linked by a disulfide bond (S–S); the bulk of the chains is exposed outside the cell; each chain carries a fatty acid, palmitate, that may help to anchor the receptor; the receptor binds two molecules of transferrin, presumably one molecule per chain. Each transferrin molecule can carry two ferric ions (Fe^{+++}).

now, of course, on their outer, or cytoplasmic, surface: the clathrin coat envelops the vesicle membrane in a fibrous network of pentagons and hexagons. (Apparently it is the polymerization of clathrin that actually forms the coated vesicle. When coated vesicles are purified free of membrane, the clathrin can be dissociated into three-armed subunit structures called triskelions. Under appropriate conditions the triskelions can aggregate to form the kind of basketlike structure that surrounds a coated vesicle.) As the coated vesicles move deeper into the cytoplasm they shed their clathrin coat and fuse with one another (or with a different kind of vesicle) to form larger, smooth-surfaced vesicles called endosomes.

Do receptors always reside in coated pits or is it only the binding of a ligand that makes them migrate there through the essentially fluid membrane? It depends on the receptor. Anderson has shown that the LDL receptors on fibroblasts (connective-tissue cells) are found primarily in coated pits even in the absence of bound LDL particles. On the other hand, receptors for transferrin, insulin and galactose-terminal glycoproteins are ordinarily distributed diffusely on the plasma membrane and seem to cluster in coated pits only when the ligand has bound and when the temperature is at least 37 degrees Celsius. A simple explanation would be that certain receptors have some property directing them to coated pits and holding them there; part of the receptor might, for example, have an affinity for a component of the pit region. Such a property might be elicited in other receptors only by the binding of its ligand, perhaps through a change in receptor conformation.

Although different ligands have different final destinations, most of them are dispatched to the lysosomes. There some 40 digestive enzymes, functioning in an acidic environment (a pH of from 4.5 to 5 in contrast to the neutral pH 7 to 7.4 of the cytoplasm), break down the ligands. The digestion products are either eliminated from the cell or (as in the case of the LDL components) exported into the cytoplasm as raw materials. It would be wasteful indeed if receptors shared the fate of their ligands, and indeed they do not. Some years ago it became clear that receptors, having delivered their ligands to the interior of the cell, are recycled to the cell surface to do it all over again. There were several lines of evidence. The making of new receptors could be blocked with specific inhibitors of protein synthesis and yet receptor-bound ligand continued to be internalized. It could be shown that the lifetime of most receptors far exceeds the lifetime of their ligand and that each receptor for transferrin, LDL or the galactose-terminal glycoproteins internalizes a ligand every 10 or 15 minutes, and keeps doing so for many hours.

Do receptors cycle continuously, like a shuttle bus, or is their cycling induced by the binding of ligand (making them more like a taxicab)? Again the answer differs for different receptors. The LDL receptors seem to keep cycling through human fibro-

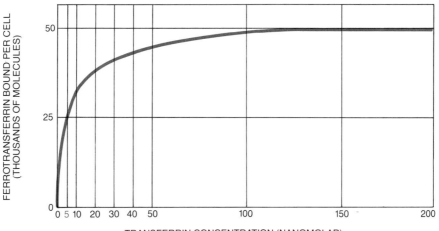

Figure 8.4 AFFINITY OF RECEPTORS for iron-loaded ferrotransferrin is calculated. Liver-tumor cells are incubated in increasing concentrations of radioactively labeled transferrin at four degrees Celsius. The ligand binds at the surface but is not internalized, and the amount bound is quantified. The maximum binding level, 50,000 molecules per cell, shows a cell has some 50,000 transferrin receptor sites on its surface. The concentration at which half of them are occupied, 5 nanomolar, is the dissociation constant, a measure of the receptor's affinity for its ligand.

blasts whether or not they have bound ligand LDL. On the other hand, there is evidence that the receptors for galactose-terminal glycoproteins are internalized only when the ligand is bound. Apparently it is the receptors that are always localized in coated pits (even in the absence of ligand) that cycle continuously, because the coated pits keep invaginating; other receptors are seen to cycle only in the presence of ligand because it is only then that they move to coated pits.

At what stage are receptors separated from their ligands, and how? Hans J. Geuze, Jan W. Slot and Ger J. A. M. Strous of the University of Utrecht, collaborating with Alan L. Schwartz and one us (Lodish), applied a novel electron-microscope technique in an effort to find out where the separation takes place (see Figure 8.5). An antibody directed against a galactose-terminal glycoprotein and one directed against the galactose receptor were made and the two antibodies were labeled with gold particles of differing size. Rats injected with the glycoproteins were sacrificed at varying times after the injection, the livers were fixed and sectioned and the sections were treated with the two antibodies, which bound to the ligand and to the receptor; the gold particles were visible as dense dots in electron micrographs.

In sections made in the early stages of endocytosis both the receptor and its ligand are seen to be closely associated with the membrane of vesicles lying just under the surface of the cell. The lumen (the fluid-filled interior) of the vesicles does not contain any free ligand. Apparently the ligand is still bound to receptors embedded in the vesicle membrane. Farther from the cell surface, however, one sees larger vesicles, and these have free ligand in the lumen; the larger the vesicle is, the more free ligand seems to be present. In these larger vesicles the receptors seem to be still associated with membrane, but they are no longer distributed randomly over the entire vesicle membrane. Rather, they either are clustered near an end of the vesicle where the vesicle seems to be adjacent to or actually fused with thin, membranous tubules, or they are in the tubules themselves.

We call this vesicle-plus-tubule structure a CURL (for "compartment of uncoupling of receptor and ligand"). It seems to be the organelle in which receptors and ligands are separated and redistributed, with the ligands accumulating in the vesicular part and the receptors accumulating in the membraneous tubules. Tissue sections made after endocytosis

has proceeded for a longer time show that the CURL vesicles eventually fuse with lysosomes, where the ligand is degraded. Before the fusion takes place, however, the tubular portions loaded with receptors have been detached from the CURL vesicles, and so the receptors escape lysosomal degradation. The tubular structures seem to serve as intermediates that somehow get the receptors back to the cell surface. Whether they are vesicles that move to the surface or are part of a fixed tubular system, something like the endoplasmic reticulum, extending to the surface is still not known.

On the surface of a cell a receptor binds its ligand tightly. What causes a receptor and its ligand to dissociate within the cell? A low pH promotes such dissociation, and a few years ago it was reported that the recycling of receptors can be inhibited by agents that raise the pH of acidic vesicles within the cell. Benjamin Tycko and Frederick R. Maxfield of the New York University Medical Center showed directly that endosomes (and presumably CURL's) are normally acidic. They studied the endocytosis in human fibroblasts of alpha-2-macroglobulin, a protein that binds proteases and is itself bound by cell-surface receptors and internalized, thereby clearing the proteases from the circulation (see Figure 8.6). Tycko and Maxfield coupled the fluorescent dye fluorescein to the alpha-2-macroglobulin. The excitation spectrum of fluorescein is sensitive to pH, that is, the intensity of its fluorescence varies with pH as well as with the wavelength of the light to which the dye is exposed.

The fibroblasts were incubated with the fluorescein-labeled ligand just long enough for the ligand to be internalized and reach prelysosomal vesicles such as endosomes or CURL's. Then the cells were chilled, illuminated at two different wavelengths and observed with an image-intensifier camera mounted on a fluorescence microscope. Calculations based on the relative intensity of fluorescence at the two excitation wavelengths showed that the pH of the vesicles containing the ligand was about 5.0.

Experiments with purified receptors and ligands show that whereas receptors bind their ligand tightly at a neutral pH, the ligands are rapidly dissociated when the pH goes below 5.5 (see Figure 8.7). It is clear that when a receptor-ligand complex enters an acidic vesicle, the ligand is separated from the receptor and becomes soluble in the vesicle's lumen; the receptor remains bound to the vesicular

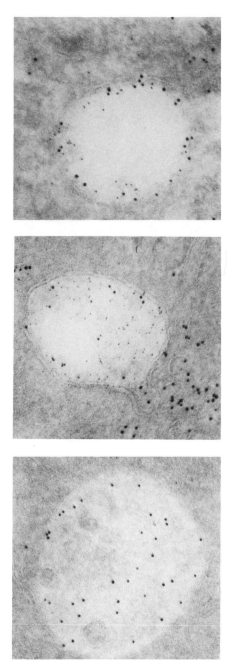

Figure 8.5 SEPARATION OF RECEPTOR AND LIGAND shown in frozen, sectioned rat liver cells that were treated sequentially with an antibody preparation specific for the ligand and then one specific for its receptor, which were indirectly labeled with gold particles of differing size. The antibodies bound to their respective antigens, whose location was revealed by the particles, which are seen as large and small dots. In an endosome (*top*), an endocytic vesicle near the cell surface, receptors (*large dots*) and ligand molecules (*small dots*) are seen together on the inner surface of the vesicle membrane. Deeper in the cell, in a CURL (*middle*), ligand molecules have been released from the receptors and are free in the interior of the vesicle; the receptors are concentrated in tubular structures attached to the lower right side of the vesicle. A micrograph made with the gold labeling reversed shows that a later-stage vesicle, a multivesicular body (*bottom*), contains ligand molecules but no receptors.

membrane and is eventually recycled. How the endosomes (or CURL's) are acidified is not yet established, but it is known that there is an enzyme in the endosomal membrane that exploits the energy stored in adenosine triphosphate (ATP) to pump protons into the lumen. An excess of protons means a low *p*H.

Unlike other ligands, ferrotransferrin is neither degraded nor stored after internalization. In-

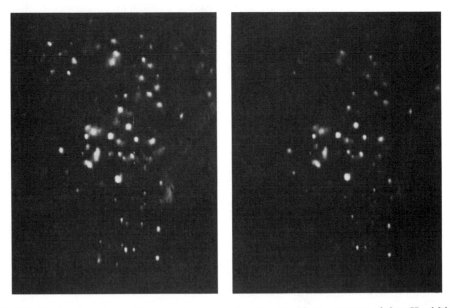

Figure 8.6 ACIDITY of endocytic vesicles was demonstrated by labeling alpha-2-macroglobulin with fluorescein. Cultured fibroblasts were incubated with the ligand just long enough for the ligand to reach endosomes or CURL's. Then the cells were illuminated with 450-nanometer violet light or with 490-nanometer blue light. The ratio of the intensities of fluorescence at the two wavelengths provides a measure of the pH within the vesicles harboring the labeled ligand. The 450-nanometer fluorescence (*left*) is somewhat brighter than the 490-nanometer fluorescence (*right*), indicating that the pH is acidic; calibration against intensity ratios measured in solutions of known pH shows that the pH in these vesicles is about 5.0.

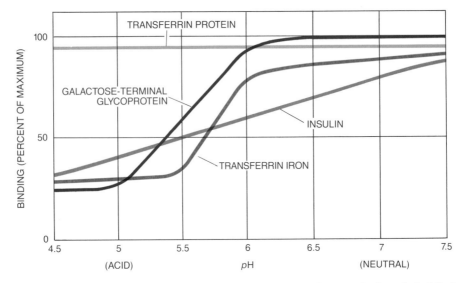

Figure 8.7 EFFECT OF ACIDIC ENVIRONMENT on the binding of various ligands differs. Whereas most ligands dissociate from their receptor as the pH is lowered below 6, the binding of transferrin to its receptor remains unaffected. When the binding of iron carried by ferrotransferrin is measured separately from that of the transferrin, the iron is seen to dissociate from the transferrin-receptor complex below pH 6 while transferrin protein remains bound to the receptor.

stead it gives up its load of iron ions, which remain inside the cell, and is itself quickly secreted from the cell as iron-free apotransferrin. Why is the transferrin protein excreted from the cell and why does it leave its iron behind? Together with Aaron Ciechanover we studied the endocytosis of transferrin in cultured human liver-tumor cells and found the answers in the differential effect of pH on the binding of ferrotransferrin and of apotransferrin to the transferrin receptor. Richard D. Klausner of the National Cancer Institute and his colleagues, working with cultured precursors of red blood cells, reached similar conclusions.

In our laboratory at the Massachusetts Institute of Technology we found that ferrotransferrin binds avidly to its receptor at the neutral pH of the extracellular environment. It is endocytosed and transferred to an acidic prelysosomal vesicle just as other ligands are. Observations some years ago had established that iron ions become dissociated from their transferrin carrier at an acidic pH (5.5 or less), and we found that to be the case even when the transferrin is bound to its receptor. Remarkably, the iron-free apotransferrin remains bound to its receptor even at pH 5, in striking contrast to such ligands as LDL, insulin and galactose-terminal glycoproteins. Indeed, the affinity of iron-free apotransferrin for its receptor at pH 5 is the same as the affinity of iron-loaded ferrotransferrin at a neutral pH 7. On the other hand, the apotransferrin has no measur-

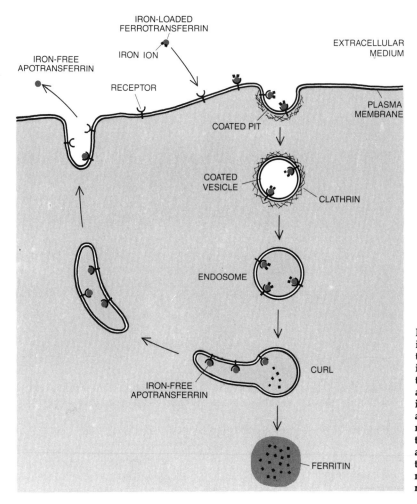

Figure 8.8 TRANSFERRIN CYCLE is traced. After endocytosis of the transferrin-receptor complex, iron is released from the transferrin in the acidic environment of the CURL and is transferred to the iron-storing protein ferritin. The iron-free apotransferrin remains bound to its receptor and is recycled with it to the cell surface. When the receptor-apotransferrin complex encounters the neutral pH of the extracellular medium, the iron-free apotransferrin is released.

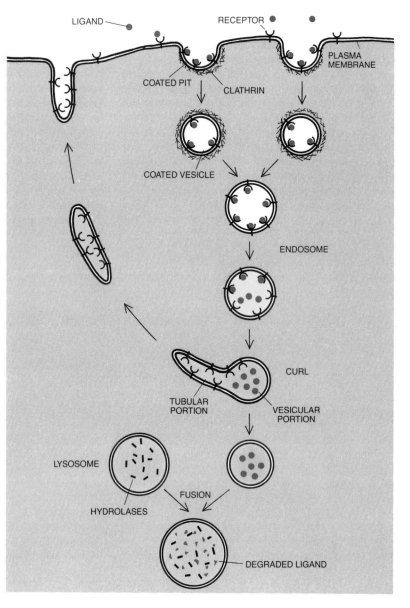

Figure 8.9 PATHWAY OF RECEP-TORS AND LIGANDS. Ligand binds to receptors diffusely and then collects in coated pits, which invaginate and are internalized as coated vesicles whose fusion gives rise to endosomes and then to a CURL. In the acadic CURL environment (*color*) ligand is dissociated from receptors. Ligand accumulates in the vesicular lumen of the CURL and the receptors are concentrated in the membrane of an attached tubular structure, which then becomes separated from the CURL. The vesicular part moves deeper into the cell and fuses with a lysosome, to which it delivers the ligand for degradation.

able affinity for its receptor at a neutral pH; if an apotransferrin-receptor complex is suddenly transferred from an acidic to a neutral environment, the bound ligand dissociates within seconds.

The course of transferrin in endocytosis can now be traced (see Figure 8.8). Ferrotransferrin, with its two iron ions, binds to the cell-surface receptor and the complex is internalized and transported to an

endosome and thence to a CURL. When the complex reaches an acidic vesicle, the iron is released. It is transported (how is not known) to ferritin, an iron-storage protein in the cytoplasmic fluid. The transferrin receptor is recycled to the cell surface, just as other receptors are after they have been freed of their ligand. The transferrin receptor is not freed of its ligand, however; it carries the iron-free apotransferrin with it to the outside of the cell, where the neutral pH of the extracellular medium causes the apotransferrin to dissociate from the receptor. The iron-free apotransferrin enters the bloodstream, ready to load more iron; the receptor takes its place in the plasma membrane, ready to bind another ferrotransferrin molecule—for which, as we mentioned above, it has high affinity in the neutral extracellular environment.

All of this happens very quickly. We can show that each of the liver-tumor cells we work with has some 150,000 transferrin-binding sites on its surface or at some stage of recycling within the cell. A cell takes up transferrin for several hours at a rate of 19,000 iron atoms, or 9,500 transferrin molecules, per minute. Dividing 150,000 by 9,500 gives the cycling time: some 16 minutes from the binding of ferrotransferrin to the secretion of apotransferrin. We have measured the time required for each step of the cycle. On the average it takes four minutes for a transferrin receptor on the surface to bind a ferrotransferrin ligand. The ferrotransferrin-receptor complex is internalized in about five minutes. It takes another seven minutes for the iron to dissociate from the transferrin and the apotransferrin-receptor complex to return to the surface, after which the apotransferrin is released from the receptor in only 16 seconds. The total elapsed time is about 16 minutes.

The entire process of endocytosis involves a succession of rapid rearrangements and fusions of biological membranes. Certain membrane proteins are selectively trapped in a coated pit, whereas others are specifically excluded. Every time a receptor-ligand complex is internalized a piece of the plasma membrane "buds off" from the surface to form a coated vesicle. The newly generated vesicles fuse, giving rise to endosomes and CURL's (see Figure 8.9). The CURL membrane is dissociated, some of it apparently fusing with the tubules that return receptors to the surface and some of it fusing with a lysosome to deliver ligands for degradation. In turn the lysosome may bud off parts of its own membrane as small vesicles that return to the cell surface and disgorge digestion products outside the cell.

Cells may internalize a large part of their plasma membrane (as much as 50 percent of it per hour), not only in receptor-mediated endocytosis but also in pinocytosis or phagocytosis. It would be wasteful for the cell to keep resynthesizing membrane components. Presumably the bits of membrane or their components are instead restored to the surface or to the organelle (such as the lysosome) to which they belong. The phospholipid composition of the plasma membrane and other membranes is somewhat different, and all biological membranes have very different complements of integral membrane proteins, of which the receptors we have been discussing constitute only one class. Somehow the plasma membrane and every other biological membrane must maintain their individuality in the face of continual invagination, fusion and recycling events.

There must be mechanisms for the continual sorting of membrane components, but nothing is yet known about such mechanisms. What accomplishes the unequal partitioning of membrane proteins so that, for example, the plasma membrane gets the permeases it should have and the lysosome gets the right proton-pumping enzyme? How are vesicles targeted from one organelle to another? Are there specific sorting signals or receptorlike molecules on the outer surface of each vesicle? More detailed understanding of receptor-mediated endocytosis will require answers to this kind of question.

III

DISORDERED CELL COMMUNICATION
AND HUMAN DISEASE

. . .

Introduction

Two aspects of intracellular signaling are germane to several of the chapters in this section. In the major signaling pathways discussed in Section II, the initiation of response is begun by the interaction of the extracellular signal with a receptor on the plasma membrane, or in the case of the retinal cell on the disc membrane. However, there is a class of extracellular messengers exemplified by steroid sterol and thyroid hormones that do not interact with surface receptors but enter the cell and bind to receptors either in the cytosol, which they translocate to the nucleus, or to receptors already in the nucleus. In either case, the major site of action of this class of receptors (when bound to their appropriate chemical messenger) is that of activating the transcription of one or usually a number of genes in the target cell.

Chapter 9, "Diseases Caused by Impaired Communication among Cells," by Edward Rubenstein, gives several examples where abnormal signaling can result in human disease because either a bacterial toxin or an immunoglobin acts as a pseudohormone and stimulates the cyclic adenosine 3'5'-monophosphate (cAMP) messenger system in one or another target cell.

Chapter 10, "Viral Alteration of Cell Function," by Michael B. A. Oldstone, focuses on the fact that a class of latent virus can greatly alter specific signaling events so that insufficient hormone or neurotransmitter is synthesized and secreted from cells that appear normal. However, as discussed in Chapter 11, "The Causes of Diabetes," by Abner Louis Hopkins, there are considerable data that indicate a viral infection in a host of the appropriate genetic makeup can also trigger the autoimmune destruction of specific signaling cells, such as the beta cells of the endocrine pancreas, resulting in insulin-dependent diabetes.

Since insulin is the crucial hormone in the maintenance of glucose homeostasis, its loss leads to severe consequences. In particular, the rate of development of atherosclerosis and aging are increased in diabetic patients, discussed in Chapter 12, "Glucose and Aging," by Anthony Cerami, Helen Vlassara and Michael Brownlee.

Chapter 13, "How LDL Receptors Influence Cholesterol and Atherosclerosis," by Michael S. Brown and Joseph L. Goldstein, brings one's attention back to the importance of surface receptor turnover in the physiology of the cell and how abnormalities in one of these receptors can lead to atherosclerosis.

The discussion of the development of atherosclerosis in Chapters 12 and 13 points up an additional level of complexity in terms of considering a disease as a disorder of cell communication. This complexity is that chronic diseases represent historical processes. As such, the nature of the dialogue between the various cells of the organism that participate in the disease process is not constant but changes as the disease progresses. Thus, as we enter the information age, and as our knowledge of inter- and intracellular communication expands, it is quite likely that our present system for classifying diseases according to an observed change in the microscopic structure of a tissue or cell—a system now more than a century old—will be replaced by one in which diseases will be classified in terms of disordered cell communication.

Diseases Caused by Impaired Communication among Cells

The activities of the cells of the body are integrated by mediator substances that interact with specific receptors. It appears that many diseases are attributable to errors in such communication.

• • •

Edward Rubenstein
March, 1980

In medicine it is both clinically useful and intellectually satisfying when the mechanism underlying a mystifying disease or group of diseases is revealed. Recently it has become apparent that a number of human diseases, among them such diverse disorders as cholera, hyperthyroidism, myasthenia gravis and certain types of diabetes, arise from a common mechanism: faulty communication among cells. The recognition of this mechanism has led to the formulation of important new concepts about a system involved in the self-government of the body. The principle components of this system are chemical signals, called mediators, and the cellular structures with which they interact, called receptors.

Cells communicate with one another by releasing mediators that travel various distances, sometimes only to an adjacent cell and at other times through long journeys in the bloodstream to other parts of the body. The messages are picked up by the receptors, which relay the information to structures within the cell where the incoming signal triggers a biochemical response.

When the mediators are steroid hormones, the information they carry travels in most instances directly to the nucleus of a cell in the target tissue, where interaction with the genetic material of the cell gives rise to the synthesis of new proteins. This process of interaction and synthesis takes a certain amount of time. When the mediators are peptide hormones or catecholamines, the information they carry leads to the modification of previously assembled proteins stored in the cytoplasm of the target cell. This process does not take much time and the biological response can be quick. In many such systems the relay signal is cyclic adenosine monophosphate (cyclic AMP), which is produced when two of three phosphate groups are cleaved from its ubiquitous precursor, adenosine triphosphate (ATP), by the enzyme adenylate cyclase (see Figure 9.1). The free end of the remaining phosphate group combines with carbon atom No. 3 in the ribose (five-carbon sugar) of the ATP to form a ring, hence the term cyclic. The concentration of cyclic AMP is increased by the level of adenylate cyclase and decreased by the level of the enzyme phosphodiesterase.

The diverse effects of cyclic AMP are attributed to

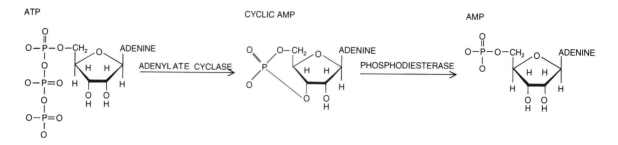

Figure 9.1 CYCLIC AMP is formed and degraded as is shown here. Its precursor, adenosine triphosphate (ATP), loses two of its three phosphate groups (P) through cleavage by the enzyme adenylate cyclase. The free end of the remaining phosphate group combines with a carbon atom in a cyclic, or ring, configuration. Cyclic AMP is inactivated by the enzyme phosphodiesterase, which opens the ring and converts the cyclic AMP into an inert form of AMP. Cyclic AMP is in many instances the second messenger that responds within a cell to a mediator, such as a hormone, that has arrived from another cell. The mediator molecules bind to specific receptors in the cell membrane.

its ability to convert certain inert proteins into functioning ones known as kinases. The kinases modulate the metabolism of the cell. They switch systems on or off by activating or deactivating enzymes through the process of phosphorylation. For the discovery of cyclic AMP and the eluciation of its central role in the metabolism of cells the Nobel Prize in medicine was awarded in 1971 to Earl W. Sutherland, Jr., of the Vanderbilt University School of Medicine.

More recent studies have indicated that receptors on the cell's outer membrane communicate with adenylate cyclase through an intermediary protein binding guanosine triphosphate (GTP). This protein not only activates the cyclase but also may diminish the affinity of the receptor for the mediator. Such a mechanism would partly explain the fact that high levels of mediators tend to reduce the response of their receptors. For example, the administration of certain adrenergic drugs to patients with asthma reduces the number of adrenergic receptors on their white blood cells. Insulin has a similar diminishing effect on its receptor, whereas estrogen increases the number of receptors for progesterone and thyroid hormone increases the number of beta-adrenergic receptors in some tissues.

The initial mediator is sometimes called a first messenger and cyclic AMP and other relay substances, which shuttle from the vicinity of the receptor to other sites in the cell, are called second messengers. The terminology is somewhat misleading, because most first messengers are not initiators in the strict sense. They are released in response to prior orders, many of which originate in the central nervous system as a consequence of processes ranging from the perception of a sensory stimulus to the dawning of an idea. The concept is a major extension of the field of neuroendocrinology, with some unexpected twists.

Mediators, Receptors and Membranes

The mediators are molecules ranging in size and complexity from amino acids (consisting of from 10 to 27 atoms) to insulin (consisting of some 925 atoms). The receptors are huge aggregates of many molecules, arranged in an assembly that gives each kind of receptor a unique structure. In general hydrophilic (water-soluble) chemical mediators cannot diffuse across the cell membrane to enter the interior of the cell; their receptors are on the cell surface. (see Figure 9.2). Hydrophobic (fat-soluble) compounds seem to diffuse across the cell membrane; their receptors are inside the cell.

One therefore sees that the cell membrane is an important component in the system of communication among cells (see Figure 9.3). The membrane consists of a double layer of phospholipid molecules, arranged so that water-soluble groups point outward toward the exterior and interior of the cell and fatty-acid chains point inward from the water-soluble groups. Interspersed among the phospholipid molecules are flat, rigid molecules of cholesterol, which are aligned with the fatty-acid chains.

Assemblages of protein molecules, many of which are thought to be receptors, float in or are

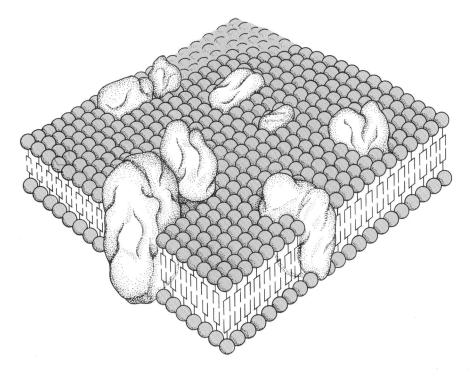

Figure 9.2 PROTEINS IN THE CELL MEMBRANE include structures that are thought to be receptors. Some of the proteins, called integral proteins, go entirely through the membrane; others, called peripheral proteins, are on the inner or outer surface of the membrane and are bound to integral proteins. Some of the proteins probably form conduits through which fat-insoluble mediators enter the cell. Fat-soluble mediators probably diffuse across the cell's fatty envelope.

attached to the membrane. Some of them extend across the thickness of the membrane; others are on the inner or outer surface. A few of the latter proteins are probably gated conduits through which fat-insoluble substances gain entrance to the interior of the cell. Fat-soluble compounds, such as androgens, estrogens, corticosteroid hormones and fat-soluble vitamins, are believed to diffuse across the fatty membrane of the cell and to encounter their specific receptors in the cytoplasm. The receptor for thyroid hormone is probably in the cell nucleus.

Water-soluble compounds, such as the protein hormone insulin, cannot passively cross the hydrophobic barrier, and therefore their receptors are situated in the cell membrane. Recent studies have shown that the membrane-bound receptors for insulin and other hormones enter the cell after they have combined with their mediators and then move in a nonrandom manner to target locations, such as the membrane of the cell nucleus, where they probably exert long-term effects by regulating the expression of genes. Perhaps such effects are initiated by the receptor rather than by the mediator, which may serve only to activate the receptor and to trigger its movement from the cell membrane to a particular site within the cell.

The interaction between the mediator molecule and its receptor is highly specific. It results from the fact that active regions on each of them have complementary contours and distributions of electric charge, enabling the mediator to come into such close contact with the receptor that the two are chemically bonded. The interaction gives rise to a sequence of biochemical reactions that produce a biological effect.

Under normal circumstances the receptor interacts only with the appropriate mediator. Hundreds of other kinds of chemical compounds may come in contact with the receptor, but they bounce off and

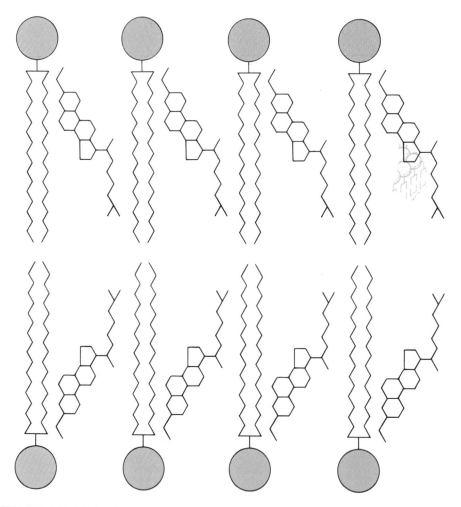

Figure 9.3 CELL MEMBRANE is a key component of the system of communication among cells. Because of its structure water-soluble and fat-soluble mediators enter the cell in different ways. The membrane consists of a double layer of phospholipid molecules, arranged so that their water-soluble groups (*colored spheres*) point both outward and inward and their chains of fatty acids point inward. Also interspersed among the phospholipid molecules are flat, rigid molecules of cholesterol, shown as pentagonal structures to the right of the fatty-acid chains.

nothing further happens. The high degree of specificity between the mediator and the receptor is demonstrated by the hormones oxytocin and vasopressin (see Figure 9.4). Each is a simple peptide compound consisting of nine amino acids; they differ from each other only in the amino acid units at the third and the eighth position. The hormones are almost identical in size, shape and the distribution of electric charge on their surface, but the subtle distinctions between them account for the fact that they have entirely different biological effects.

Oxytocin activates receptors in the smooth-muscle cells of the wall of the uterus, initiating the contraction of the uterus in labor. Vasopressin activates receptors in epithelial cells that line the collecting tubules of the kidney, causing the tubules to become more permeable and thereby allowing the kidney to reabsorb large amounts of water. (The result is a reduction in the volume of urine.) Vasopressin does not act on the uterus and oxytocin does not act on the kidney.

Although that specificity is exceedingly high, it is

Figure 9.4 SPECIFICITY OF A CELL-COMMUNICA-TION SYSTEM is illustrated by the hormones oxytocin and vasopressin. Each of these hormones consists of nine amino acids. They are identical except at positions 3 and 8, and yet they have entirely different functions and presumably have different receptors. Oxytocin acts on the uterus, vasopressin on the kidney.

not complete. A receptor may on occasion react with an inappropriate compound if the electron clouds of the compound resemble those on the active region of the proper mediator molecule. The receptor then responds to the wrong signal. An example is the action of the drug carbachol on synapses in the autonomic nervous system, which controls such largely involuntary functions as blood pressure and contractions of the intestine.

Where two neurons, or nerve cells, meet at a synapse they are separated by a space filled with fluid. At the terminal of the presynaptic neuron are membrane-enclosed vesicles that contain the neurotransmitter acetylcholine. When a nerve impulse reaches the nerve ending, the molecules of acetylcholine are released in a burst; they diffuse across the narrow gap between the cells and plug into the socketlike cavities of the acetylcholine receptors that cover the surface of the postsynaptic membrane.

The drug carbachol is similar enough to acetycholine to substitute for it and so to act as a synthetic neurotransmitter. The receptor responds to a false signal, which in this instance is a compound employed as a drug. In other instances faulty communication is the result of a physiological defect involving either the mediator or the receptor.

These comparatively simple new concepts of cell physiology are revolutionizing concepts of many common disease processes. An example is cholera, one of mankind's worst scourges. Although cholera has for years been known to be caused by the ingestion of large numbers of the bacillus Vibrio cholerae, the disease itself is not an infection. The microorganism does not invade the tissues; it merely colonizes the intestinal canal for a few days. It cannot penetrate or burrow between the cells that line the gut, it is incapable of gaining access to the lymph channels and it does not enter the bloodstream. There is no microscopic evidence that it damages any tissue. The entire disease process, which kills half of its untreated victims, is the result of the fact that cells in the intestine respond to a chemical substance released by the bacteria as if it were a normal regulatory signal.

Error in Cholera

The principal functions of the small intestine are the digestion and absorption of food. These processes are carried out by the degradation of proteins, carbohydrates and fats by specific enzymes secreted in the small intestine and the pancreas. The enzymes work best in an alkaline fluid. Hence when food is delivered from the stomach to the small intestine, a chemical signal (of uncertain origin) interacts with receptors on the cells of the small intestine and stimulates an adenylate cyclase system that leads to the pumping of about two liters of alkaline fluid into the intestinal canal. The fluid provides a medium for the digestion of food and then is reabsorbed farther along in the small intestine and in the colon.

The work of many investigators has led to the current recognition that a dysfunction of receptors is the key to cholera (see Figure 9.5). W. E. van Heyningen and his colleagues at the University of Oxford and the Johns Hopkins University School of Medicine demonstrated that the toxin produced by cholera bacteria binds to receptor sites on cells lining the small intestine. Two groups at the Harvard Medical School (Daniel V. Kimberg, Michael Field and their associates and Geoffrey W. Sharp and Sixtus Hynie) established that the toxin of cholera bacteria causes the secretion of fluid by the small intestine by increasing the activity of adenylate cyclase in cells of the intestinal mucosa.

The molecule of the cholera toxin consists of two subunits. One binds with a patch on the surface of the receptor. The other, which is held in place by

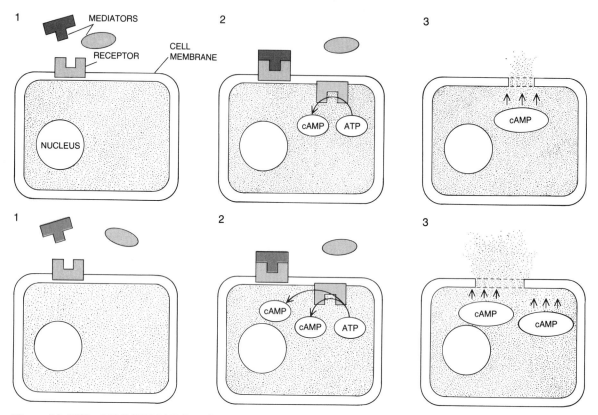

Figure 9.5 CELL COMMUNICATION of a normal type (*top*) and of an abnormal one in cholera (*bottom*) is depicted for a cell of the small intestine. In normal communication a receptor encounters many kinds of mediators flowing past it in the bloodstream. It responds only to a specific mediator released when food is delivered to the small intestine. The interaction of the mediator and the receptor activates a relay system consisting of cAMP, which is manufactured from ATP. The cAMP activates an enzyme that causes the cell to discharge into the intestinal canal an alkaline fluid that enhances the activity of the digestive enzymes. Later the digestive tract reabsorbs the fluid. In cholera the normal mediator is mimicked by a toxin released by the cholera bacillus. When the toxin binds to the receptor, the cyclic-AMP system is overstimulated, causing the cell to release large amounts of fluid into the intestinal canal. When many such cells are overstimulated, far more fluid is produced than the digestive tract can reabsorb.

the first, stimulates the adenylate cyclase system into such overactivity that the cells pump from 20 to 30 liters of water into the small intestine. Because fluid cannot be reabsorbed at that rate massive amounts of it are lost through vomiting and diarrhea. These huge losses of liquid account for the many deaths from cholera. The threat to life can be countered by administering adequate amounts of fluid to the patient. The antibiotic tetracycline also helps, but only to the extent that it shortens the duration of the disease and reduces the likelihood that the patient will become a carrier.

Other Diseases

Graves' disease, a common cause of hyperthyroidism, is characterized by an enlargement of the thyroid gland and a massive overproduction of thyroid hormone. The excess output of the hormone has many adverse effects, largely because of hypermetabolism that is secondary to the inefficient expenditure of cell energy. The patient is likely to exhibit nervousness, tremor, a fast pulse rate, sweating, intolerance of heat, loss of weight and other symptoms.

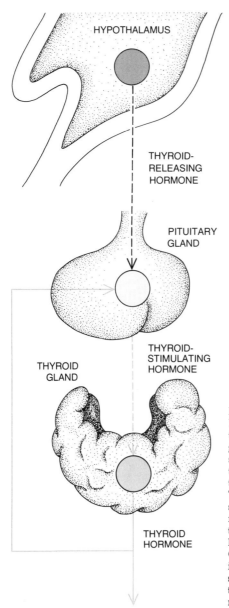

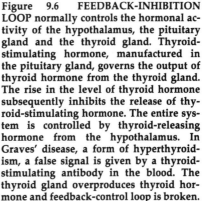

Figure 9.6 FEEDBACK-INHIBITION LOOP normally controls the hormonal activity of the hypothalamus, the pituitary gland and the thyroid gland. Thyroid-stimulating hormone, manufactured in the pituitary gland, governs the output of thyroid hormone from the thyroid gland. The rise in the level of thyroid hormone subsequently inhibits the release of thyroid-stimulating hormone. The entire system is controlled by thyroid-releasing hormone from the hypothalamus. In Graves' disease, a form of hyperthyroidism, a false signal is given by a thyroid-stimulating antibody in the blood. The thyroid gland overproduces thyroid hormone and feedback-control loop is broken.

Ordinarily thyroid-stimulating hormone (TSH), which is manufactured in the pituitary gland, governs the activity of the cells in the thyroid that produce thyroid hormone. Thyroid-stimulating hormone binds with receptors on the cells and then activates adenylate cyclase. The increased levels of cyclic AMP lead to the synthesis of thyroid hormone and its release into the circulation. Thyroid hormone in turn inhibits the release of thyroid-stimulating hormone by the pituitary gland. Hence a simple feedback-inhibition loop regulates the function of the thyroid gland (see Figure 9.6). The base-line activity of the system is governed by TSH releasing hormone (TRH), which is manufactured in cells of the hypothalamus.

The blood plasma of patients with Graves' disease contains a thyroid-stimulating antibody. S. Qasim Mehdi and Joseph P. Kriss of the Stanford University School of Medicine have shown that this immunoglobulin interacts with receptors of thyroid-stimulating hormone in the cells of the thyroid gland, competes with the pituitary hormone and continuously activates the receptors. What triggers the production of the antibody remains unknown. Prior damage to thyroid tissue appears to be important in some instances; a genetic predisposition may also play a role.

In any event the thyroid gland is now driven by a stimulating antibody rather than by the normal regulatory hormone. Moreover, the feedback-inhibition loop has been broken. Overproduction and oversecretion of thyroid hormone ensue, and the patient develops all the serious clinical abnormalities associated with hyperthyroidism. The entire disease process appears to be the consequence of the functional similarity between a part of the antibody molecule and a part of the TSH molecule.

A somewhat similar mechanism is responsible for myasthenia gravis, an often serious disorder in which the patient develops a profound muscle weakness. The repetitive use of a muscle intensifies the symptoms. Jaw muscles become tired after prolonged chewing; the upper eyelids droop after they have been kept open for a sustained period; the voice falters in long conversations.

It is found that most of these patients have a circulating antibody that not only combines with receptors of acetylcholine in the end plates of the motor nerves but also destroys the receptors. Some patients show clear evidence of abnormal function of the thymus gland, which is responsible for the production of one class of immune cells. A passive transfer of the antibody across the placenta to the fetal circulation accounts for the occurrence of the disease in the offspring of affected mothers. Within weeks after birth the antibodies are degraded in the infant's tissues and the infant recovers spontaneously.

A series of ground-breaking experiments by Michael S. Brown and Joseph L. Goldstein of the University of Texas Health Science Center at Dallas has established that a dangerous form of elevated blood cholesterol is another disease of receptors.

This disorder, which is inherited and transmitted to half of the offspring, results in a marked elevation of the level of cholesterol in the blood, premature coronary disease and deposits of cholesterol in the skin and certain tendons.

Cells require cholesterol to replenish cholesterol molecules in the cell membrane that have been lost through wear and tear. The liver synthesizes cholesterol from precursors and releases it into the blood as part of a low-density lipoprotein (LDL) molecule, which is removed from the blood by specific receptors on the cell membrane. When a cell is in need of supplemental supplies of cholesterol, there is an increase in the number of LDL receptors, which then intercept molecules of LDL that are flowing past the cell in the interstitial fluid. When the needs of the cell have been met, there is a decrease in the number of LDL receptors, allowing the LDL cholesterol to float past and reach other cells.

Patients with familial hypercholesterolemia, the name of the inherited disorder, are unable to produce adequate numbers of LDL receptors on cell membranes (see Figure 9.7). The blood level of LDL (and therefore of cholesterol) increases, and the damaging effects of excessive cholesterol in the blood ensue. People who are homozygous for this disorder (have inherited the defective gene from both parents) have virtually no LDL receptors. They develop exceedingly high levels of LDL cholesterol in the plasma and often suffer fatal coronary disease in childhood or adolescence. The disorder can be detected early in fetal life by analyzing amniotic fluid. The parents can then be counseled about the difficulties that lie ahead. Another form of the disease has been recognized in which the number of

Figure 9.7 A CELL-COMMUNICATION MECHANISM is evident in these micrographs. A pit in the membrane of a human fibroblast (*top*) is lined with receptors for low-density lipoprotein, which cells need to maintain the structure of their membranes. The black spots above the surface of the pit are particles of low-density lipoprotein bound to the protein ferritin, which is dense in electrons and so enhances the visibility of the particles. A pit closed over the particles forms a vesicle that is delivered to the interior of the cell (*bottom*), where the cholesterol is put to use. A failure of this mechanism is responsible for familial hypercholesterolemia. The victim therefore suffers the pathological effects of excess cholesterol in the blood.

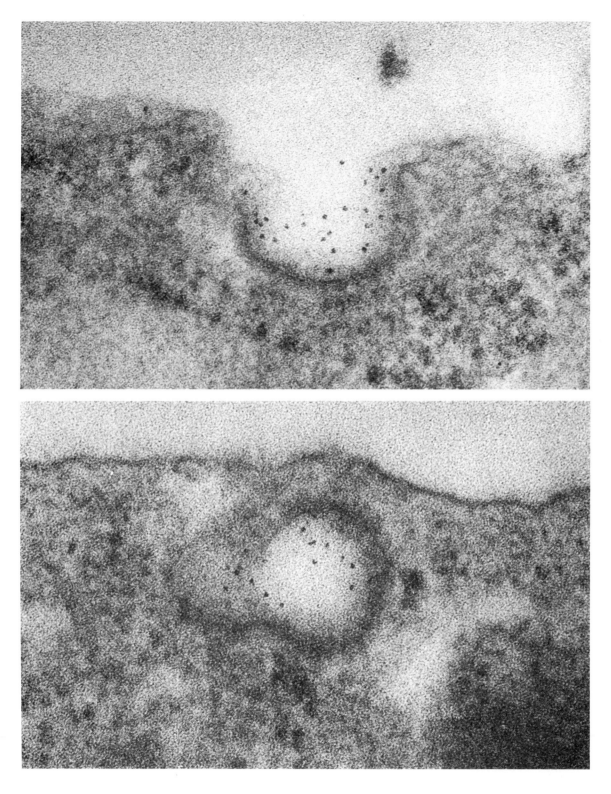

LDL receptors is normal but the process by which cholesterol-laden receptors are moved from the cell's surface to its interior is impaired.

Diabetes and Insulin

A number of types of diabetes mellitus, some of which are common, are now thought to be associated with the defective function of receptors. The tale begins with the interaction of blood glucose and insulin. Blood glucose, which is derived either from the digestive breakdown and absorption of dietary carbohydrates or from stores or manufacture in the liver, is an important source of energy and of building-block molecules for cells. It is essential for the minute-to-minute survival of neurons in the central nervous system. Glucose cannot gain entrance through the membranes of muscle and fat cells without the mediation of insulin. As the concentration of glucose in the blood rises after a meal, the level of insulin in the blood (manufactured and released by the beta cells of the pancreas) increases similarly. The insulin activates specific receptors in cell membranes (see Figure 9.8). The receptors probably provide a passageway through which glucose and other substances can enter.

It has recently been shown that soon after insulin combines with receptors the receptors move from the membrane to target structures within the cell. The movement is highly directed. Eventually many of the insulin-receptor complexes take up positions on the surface of the membrane of the cell nucleus, where they presumably cause some of the long-term biosynthetic effects associated with the action of insulin.

Insulin also plays a key role in regulating the metabolism of carbohydrate. When the level of insulin in the blood plasma is high, the liver stores glucose in the form of glycogen. When the level of insulin in the plasma is low, glycogen is broken down and glucose is released into the bloodstream.

How does impaired communication among cells cause diabetes? One form of the disease, called juvenile, brittle or insulin-dependent diabetes, is probably the result of a viral infection of the beta cells of the pancreas. Either the viral invasion itself or the antibodies produced against the virus destroy the beta cells and permanently halt the manufacture of insulin. In people with such a severe deficiency of insulin the liver releases large amounts of glucose that cannot be taken up by muscle and fat cells. The concentration of glucose in the blood rises sharply. The administration of insulin by injection stops the overproduction of glucose by the liver and activates the receptors, so that the sugar can gain entry to the cells, and thereby reverses many of the metabolic derangements associated with this type of diabetes.

Gerald M. Reaven and Jerrold M. Olefsky of the Stanford School of Medicine have shown that there are reduced numbers of insulin receptors on the cells of several types of tissue in a common form of diabetes that appears primarily in middle-aged people. In some of these people the output of insulin seems to be excessive. When the disease occurs in association with obesity, an improvement can be brought about by a reduction of weight. The concentration of insulin receptors on cell membranes appears to increase as the excess intake of calories is corrected. Whether or not the improvement in the diabetes is attributable to the increased number of insulin receptors, however, has not yet been established. Another form of diabetes has been attributed to a structural abnormality of insulin.

Jesse Roth and his co-workers at the National Institute of Arthritis, Metabolism, and Digestive Diseases have shown that obesity in general is associated with a decrease in the number of insulin receptors and that the number returns to normal when the calorie intake is made negative, that is, less than the person needs to maintain a given weight level. The same workers have demonstrated a decreased affinity of receptors for insulin in the rare disease known as ataxia telangiectasia. (The victims of this genetic disease suffer from uncoordinated muscle contractions and recurring infections owing to an impaired immune response.) Such studies have provided strong evidence that insulin receptors play a role in certain disorders of glucose metabolism.

In a related study at the National Institutes of Health, Jana B. Havrankova, Michael J. Brownstein and Roth found insulin receptors widely distributed throughout the brain of the rat. Although it is not clear whether the receptors are on nerve cells or on the cells of supporting connective tissue and blood vessels, their high concentration in the pathways controlling feeding activity is an intriguing discovery. It has long been known that insulin is not required for the entry of glucose into the neurons of the central nervous system. Does insulin then play some other role there, such as influencing the appetite? Does this finding provide a clue to the link between overeating and diabetes?

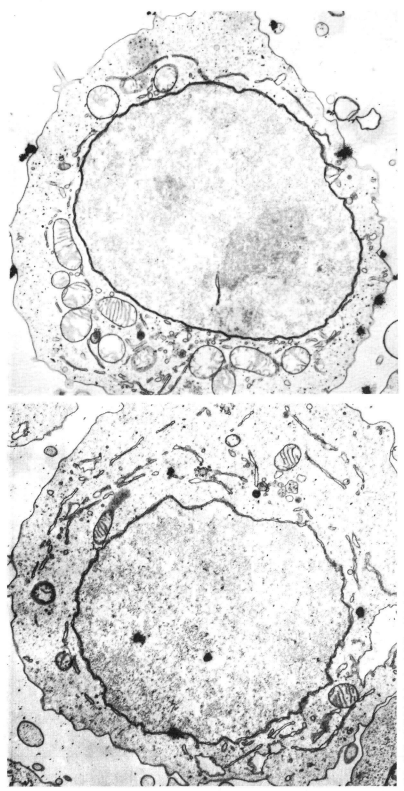

Figure 9.8 MOVEMENT OF MEDI-ATOR, in this case insulin, is shown in these micrographs of cultured human lymphocytes, or lymph cells. The cells were incubated with radioactively labeled insulin and then autoradiographs were made; the insulin appears as black spots of irregular shape. In the upper micrograph, made after 30 seconds, all the insulin is around the perimeter of the cell. In the lower micrograph, made after 30 minutes, the insulin has penetrated the cell and some of it appears to be inside the nucleus. Presumably the insulin was bound to receptors that are specific for it and that cause it to be moved inside the cell.

Substances in the Brain

A large number of other chemical substances are concentrated in brain cells but do not serve as neurotransmitters. Many of them appear to be modulators of the activity of nerve cells, scaling up or down the response at a synapse to the arrival of various electrical inputs (see Figure 9.9). The compounds include sex hormones, prostaglandins, adrenal glucocorticoids and endogenous opioids. (A remarkable endogenous opioid, a brain peptide with a high specificity for its receptor, was recently discovered by Avram Goldstein of the Stanford School of Medicine and the Addiction Research Foundation and named dynorphin by him. It is 200 times more potent than morphine and 50 times more potent than the previously identified beta endorphin. Its function may be related to how the brain regulates the tolerance of pain.) Some of the compounds appear to interact with biochemical processes directly, whereas others function through specific receptors.

Studies in rodents have shown that gonadal hormones not only influence urges and drives but also play an important role during critical periods of early development in determining the sexual differentiation of the brain. It is likely that in such animals sex steroids influence the development of neuronal pathways and the function of neurons.

The relative importance of hormonal and environmental influences on gender development in

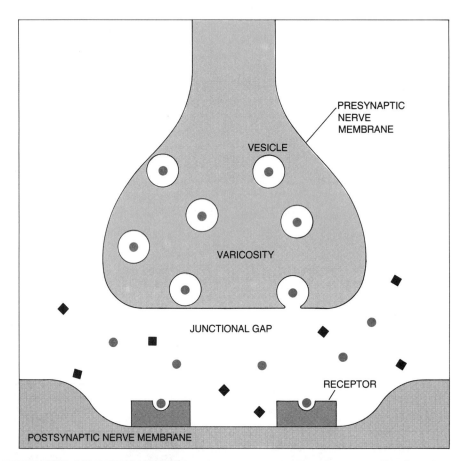

Figure 9.9 NERVE-CELL COMMUNICATION takes place at synaptic junctions. The nerve cell ends in varicosities, or swellings, that contain packets of a transmitter substance, in this case acetylcholine (*color*). On the arrival of a nerve impulse the acetylcholine is released from the vesicles into the gap between the nerve cell and another cell. Receptors on the membrane of the latter cell bind only the acetylcholine and not any of the other chemical substances (*black squares*) nearby, and the signal is thereby transferred.

human beings is the subject of controversy. Studies by Julianne Imperato-McGinley and her co-workers at the Cornell University Medical College and at the National University Pedro Henríquez Ureña in the Dominican Republic of a rare genetic form of pseudohermaphroditism found among villagers in the Dominican Republic appear to indicate that androgens play an important role in the formation of male gender identity. During or shortly after puberty almost all the affected children, who have male chromosomes and levels of testosterone in the blood that are typical of boys but who have been raised as girls because their external genitalia appear to be female, change their gender identity and gender roles. Clearly more information is needed about the contributions of chromosomes, gonads, hormones and receptors (including sex-hormone receptors in the brain) to resolve some of the complex issues surrounding gender dysfunction.

Indirect evidence of impaired chemical communication among cells has been found in a number of abnormal behavioral states. For example, functional deficiencies of the neurotransmitters serotonin and norepinephrine appear to be involved in some forms of depression, which can be precipitated or exacerbated by the administration of drugs that deplete the neuronal supply of those compounds. Recently it has been found that some antidepressant drugs act by blocking certain histamine receptors in the brain. Many of the manifestations of schizophrenic states appear to be the expression of excessive activity of dopamine in certain nerve pathways. The symptoms are intensified by the dopaminelike effects of amphetamines and relieved by a large number of antipsychotic agents, all of which are known to block neurotransmission involving dopamine.

Several neurological diseases are now viewed as probable examples of neurotransmitter dysfunction. Consider Parkinson's disease, which is the result of damage to certain lower centers in the brain caused by encephalitis, arteriosclerosis, poisons or metabolic defects. Its symptoms are disturbances in gait or posture characterized by the failure of the arms to swing during walking and by small, shuffling and slow footsteps, interrupted by sudden, spontaneous, hurried movements that propel the body forward in a bent-over position. Spasticity of the muscles of the eyelids, cheeks and mouth irons out the facial expression of emotions, which are feebly communicated even when they are deeply felt. Rhythmical tremors often appear, particularly when the affected person is at rest.

Chemical analysis of the brain tissues involved in Parkinsonism has revealed a marked depletion of dopamine and its metabolites. Because dopamine cannot penetrate the walls of the capillaries in the brain a precursor, L-DOPA, is employed therapeutically to replenish the supplies of the neurotransmitter. The treatment has been further refined by the addition of a drug that inhibits the breakdown of L-DOPA in the peripheral tissues. In some patients the regimen is of little benefit; in others it is remarkably effective, probably because the underlying defect is limited to functions for which dopamine is the mediator.

Another possible neurotransmitter disease is Huntington's chorea, a devastating hereditary disorder of the central nervous system that causes a progressive dementia and ceaseless jerky and writhing movements. The disease is not clinically evident until the afflicted person is well into adult life, and therefore it is frequently passed to individuals in the next generation, half of whom will be affected. Victims often go through years of dread and depression, and a frequent cause of death is suicide.

Certain areas of the brain of patients dying of this disease show large losses of neurons that contain the neurotransmitter gamma-aminobutyric acid (GABA). Recent studies suggest that abnormalities of phospholipids in the membrane of the neuron prevent GABA from gaining access to its receptor. A search for an analogue drug capable of reaching brain cells and activating GABA receptors is under way. (GABA does not cross the physiological barrier between the blood and the brain.)

When specimens of primary breast cancer are obtained by biopsy or in the course of surgery, about two-thirds of them show the presence of receptors that bind estrogen. A reliable clinical assay for the detection of such receptors is now widely available. (Tumors from postmenopausal women are more likely to contain estrogen-binding receptors than tumors from premenopausal women, which is strange because the older women are no longer producing estrogen.) The majority of the breast cancers found to contain these receptors respond to hormonal manipulations, and the test is therefore useful in selecting patients for whom the treatment is likely to be effective.

At least two kinds of receptors respond to histamine and both are important in drug therapy (see Figure 9.10). What are collectively called H_1 receptors are found in the respiratory tract; they mediate some of the vascular, smooth-muscle and secretory changes associated with hay fever and asthma. H_2

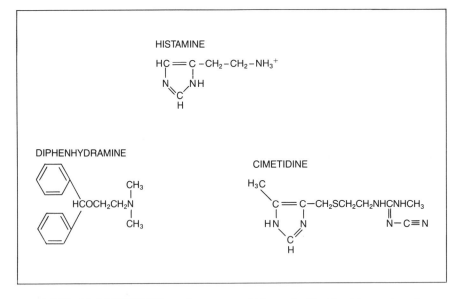

Figure 9.10 HISTAMINE AS MESSENGER evokes responses from two different kinds of receptors. The H_1 receptors are in the respiratory tract and mediate the events of hay fever and asthma; the H_2 receptors are in the stomach and mediate the secretion of hydrochloric acid. Curiously, the molecular structure of diphenhydramine, which acts by blocking H_1 receptors, bears no resemblance to that of histamine, whereas the structure of cimetidine and related compounds, which block H_2 receptors, is similar to that of histamine. Ordinarily a substance that mimics the effects of a mediator resembles the mediator itself.

receptors are found mainly in the stomach; they mediate the gastric secretion of hydrochloric acid.

The conventional antihistamine drugs block H_1 receptors and have long served in the treatment of allergic disorders. Their chemical configuration bears no resemblance to that of histamine. Such agents have no effect on H_2 receptors. Compounds of a new class, related to the drug cimetidine, powerfully block H_2 receptors and are useful in reducing the secretion of acid in certain peptic ulcers. Cimetidine and similar compounds resemble histamine structurally and so serve as substitute mediators.

Another important new class of drugs blocks beta-adrenergic receptors in the heart and blood vessels. These agents, of which propranolol is at present the one most widely administered in the U.S., blunt the effects of adrenergic stimulation. They reduce the work of the heart during physical or psychic stress and so prevent attacks of chest pain in patients whose coronary blood flow is inadequate. The drugs are also effective in preventing certain types of heart-rhythm disorder and in treating some forms of high blood pressure.

The concepts of mediators, receptors and disordered communication among cells are new and are therefore difficult to put in proper perspective. Nevertheless, the ideas I have described (having mentioned only a few of the many investigators who have made important contributions to this field) represent important advances in biology and medicine. It is likely future research will reveal that many diseases are caused by the overproduction or underproduction of a mediator or by the faulty structure of receptors or of the organelles with which the receptors interact.

Viral Alteration of Cell Function

Certain viruses interfere subtly with a cell's ability to produce specific hormones and neurotransmitters. Persistent infections by such viruses may underlie a multitude of glandular and organic disorders.

. . .

Michael B. A. Oldstone
August, 1989

"It has been well said that a virus is a piece of bad news wrapped up in protein," wrote Peter and Jane Medawar a few years back. The description is apt: when a virus infects a cell, the viral genes—the "bad news"—can, and often do, disturb the cell's normal activities. The injured cells may then die or malfunction, which gives rise to disease in the organism. Indeed, it was through the visible devastation viruses wrought in plants and animals that science first became aware of their existence.

Today a pathologist or virologist who suspects a viral infection looks with a microscope for infected cells leaking to death, their membranes riddled with escaping virus particles, as lymphocytes and other immune cells close in like riot police to contain the infection. Such immune cells themselves can damage tissues. If one finds these indicators, one then searches the patient's blood, urine and tissues for viral antibodies, genes and proteins in the hope of identifying the virus itself. In the absence of these hallmarks, the problem is usually assumed not to be viral in origin.

Within the past decade, however, a number of investigators, including my group at the Scripps Clinic, began to turn up evidence suggesting that the situation may not always be so clear-cut. We have found viruses that reside in cells and yet do not produce the classic hallmarks of viral infection: the viruses do not kill the cells, and at the same time they do not elicit an effective immune response. These tactics enable the viruses to establish a long-term presence within cells, where they can have a subtle and persistent effect—often by altering a specialized function of the cell, such as the production or secretion of a hormone. Such "luxury" functions are not essential to the cell's survival but may be vital to the health of the organism. There is now increasing evidence that this insidious mode of viral activity may underlie many human illnesses, such as certain kinds of growth retardation, diabetes, neuropsychiatric disorders, autoimmune disease and heart disease, that have not been suspected to have an infectious cause.

That this should be so is not surprising if one pauses to reflect on the singular nature of viruses. Viruses are genetic parasites, unable to sustain an independent life; they must infect living cells and exploit the cellular machinery in order to replicate. Within this limitation, however, viruses have evolved a diversity of survival strategies—and equally diverse ways in which they cause disease (see Figure 10.1). There are, first of all, the agents — such as the poliomyelitis virus and the common cold virus — that can cause acute infection and rapid cell death. There are latent viruses, such as the notorious herpes simplex, which can lie dormant and undetected in cells for long periods before flaring up, painfully. Another type of chronic infection is caused by "slow viruses," which build up gradually over a long period before giving rise to illness; slow viruses have been implicated in two rare human diseases, kuru and Creutzfeldt-Jakob, both marked by progressive dementia.

Evidence of yet another mode of persistent viral activity came from investigations by my group and others of lymphocytic choriomeningitis virus (LCMV), a disease that is endemic in certain wild mouse populations. Studied intensively since the 1930's, LCMV has been a Rosetta stone for deciphering the mechanisms of viral infection and immune response. It was known that the virus could persistently infect cells without killing them. Such studies had been conducted on fibroblasts, however — a type of cell that does not have particularly interesting specialized functions. My colleagues and I wondered what effect LCMV would have on the specialized functions of differentiated cells.

With this in mind, we decided to infect neuroblastoma cells, which produce enzymes that make and break down the neurotransmitter acetylcholine. We noted that the infected cells developed abnormalities in the synthesis and degradation of acetylcholine. Yet the cells continued to grow normally and to produce normal levels of total DNA, RNA, protein and vital enzymes. Under the microscope the infected cells were indistinguishable from uninfected ones (see Figure 10.2).

Howard Holtzer and his colleagues at the University of Pennsylvania found a corresponding situation in differentiating chick cells infected by a variant of the Rous sarcoma virus. The virus variant replicated only at certain temperatures, so that one could specify the effect of the virus on the cells by shifting the temperature. At temperatures unsuitable for viral replication the cells continued to produce the normal complement of their specialized products. When the temperature was shifted to permit viral replication, however, the cells ceased to manufacture their characteristic products. The morphology of the cells also changed, but they continued to carry on the activities they depend on to survive.

Subsequently a number of investigators found that other viruses also interfered with specialized functions of cells without disturbing their vital func-

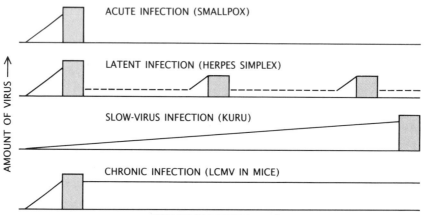

Figure 10.1 AFTER INITIAL VIRAL INFECTION there is an incubation period during which the amount of detectable infectious virus (*solid line*) increases until it brings on disease symptoms (*box*). Latent viruses may cause an initial acute infection and then lapse into a quiescent phase during which the virus is not readily detectable (*broken line*); the virus can later flare up and cause a recurrence of disease. Persistent viruses can initiate chronic infections in which the virus is detectable but does not elicit a sufficient immune response to clear the virus. Findings suggest that such chronic infections can disturb cell function and give rise to disease in the absence of classic indicators.

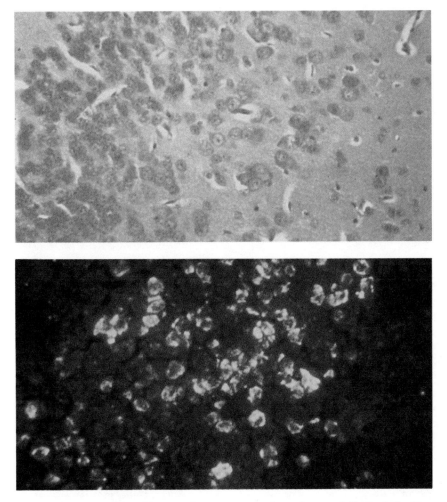

Figure 10.2 NEURONS from a mouse with a persistent lymphocytic choriomeningitis virus (LCMV) infection appear outwardly normal in an ordinary photomicrograph (*top*). The presence of the virus is revealed in a fluorescence photomicrograph (*bottom*); the fluorescent dye is delivered to infected cells by an antibody to an LCMV protein.

tions. Certain human and animal viruses were shown to infect and thereby alter the specialized functions of neurons (nerve cells), glia (a type of brain cell) and lymphocytes. It is interesting that the various viruses observed to infect lymphocytes were shown both to prevent the synthesis of *B* cells of immunoglobulin (antibody) molecules and to interfere with the ability of cytotoxic *T* cells to destroy infected cells.

These novel observations were made in the test tube. We wondered whether they would also pertain to whole animals. The first affirmative evidence came from our studies of a particular strain of mice (called C3H) that is chronically infected with LCMV. These mice had retarded growth and hypoglycemia (low blood glucose). Why should this be? Since both effects are regulated by growth hormone, we decided to examine the cells in the pituitary gland that make and secrete growth hormone. We found that the virus had invaded the anterior lobe of the pituitary and, what is more, was replicating preferentially in cells producing growth hormone. The result was that the animals produced 50

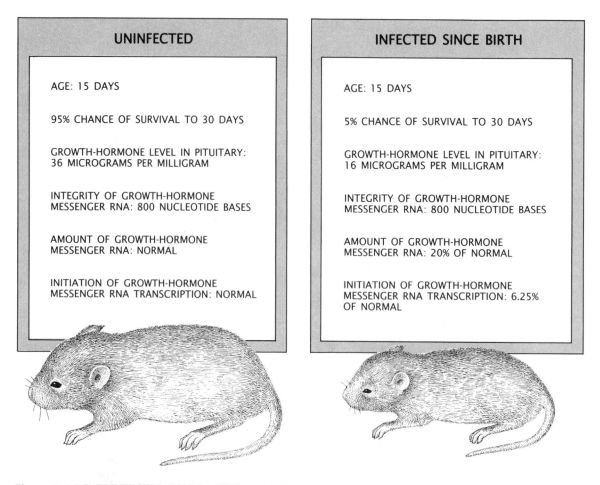

UNINFECTED	INFECTED SINCE BIRTH
AGE: 15 DAYS	AGE: 15 DAYS
95% CHANCE OF SURVIVAL TO 30 DAYS	5% CHANCE OF SURVIVAL TO 30 DAYS
GROWTH-HORMONE LEVEL IN PITUITARY: 36 MICROGRAMS PER MILLIGRAM	GROWTH-HORMONE LEVEL IN PITUITARY: 16 MICROGRAMS PER MILLIGRAM
INTEGRITY OF GROWTH-HORMONE MESSENGER RNA: 800 NUCLEOTIDE BASES	INTEGRITY OF GROWTH-HORMONE MESSENGER RNA: 800 NUCLEOTIDE BASES
AMOUNT OF GROWTH-HORMONE MESSENGER RNA: NORMAL	AMOUNT OF GROWTH-HORMONE MESSENGER RNA: 20% OF NORMAL
INITIATION OF GROWTH-HORMONE MESSENGER RNA TRANSCRIPTION: NORMAL	INITIATION OF GROWTH-HORMONE MESSENGER RNA TRANSCRIPTION: 6.25% OF NORMAL

Figure 10.3 LCMV-INFECTED MOUSE (C3H strain) is compared with its uninfected littermate. The infected mouse's stunted growth rate and low blood glucose level were caused by a shortage of growth hormone. The shortage was traced to a sharp reduction in the initiation of transcription of the messenger RNA coding for growth hormone. The infection had no effect on the messenger RNA coding for other proteins, such as actin and thyroid-stimulating protein. The infected mouse regained normal growth and glucose metabolism after receiving a transplant of healthy, growth-hormone-producing pituitary cells.

percent less growth hormone than is normal (see Figure 10.3).

When we examined the infected cells under the microscope, we saw no evidence of cell injury or inflammation. Were the stunted growth and aberrant glucose metabolism caused by a malfunction of these cells or by some other, undiscovered defect? To find out, we transplanted healthy, uninfected growth-hormone-producing cells into the infected mice to see whether the transplanted cells would restore normal growth-hormone levels. The operation was a success: the growth rate and glucose metabolism of the mice returned to normal after the transplants. We had proved that infection by LCMV had interfered with the ability of the pituitary cells to synthesize growth hormone. Although the telltale footprints of viral infection—cell injury and inflammation—were absent, the virus had disturbed homeostasis and caused disease.

Having determined that LCMV infection leads to a growth-hormone deficiency (see Figure 10.4), we wanted to learn the underlying biochemical cause of the deficiency. We first looked for a defect in the protein structure of the hormone produced by in-

Figure 10.4 DIMINUTIVE SIZE of the midget Major White, shown held aloft by the giant James Tarver in this 1926 photograph, was caused by a deficiency of growth hormone. The disorder is genetic in some cases, but in others the cause is unknown. Michael Oldstone's discovery of a persistent virus that interferes with the production of growth hormone in mice suggests that it would be worth examining growth-hormone deficiency and other hormonal and neurotransmitter disorders for possible viral origins.

fected mice by comparing it with that of the growth hormone produced by uninfected mice. Since it is difficult to carry out a sequence-by-sequence comparisons, we chopped up the protein into its component peptides and compared them instead. The result showed that the growth hormone in both infected and uninfected mice had the same components. Because protein is made according to information encoded on a strand of messenger RNA, we next examined the growth-hormone messenger RNA and found it to be the same length in both infected and uninfected mice. It was therefore unlikely that the defect lay in the protein or RNA structure of the hormone itself.

We next wondered whether infected cells were simply unable to manufacture enough of the hormone. Such a shortage could be caused by a deficiency in the amount of messenger RNA. This supposition turned out to be correct when we analyzed the RNA in the pituitary cells: we found that the level of growth-hormone messenger RNA in infected mice was only one fifth of the level in age- and sex-matched uninfected controls. The information in messenger RNA is transcribed from the DNA of a gene, and so we theorized that the flaw might be in the transcription of growth-hormone DNA into messenger RNA. And indeed, such was the case: transcription of the gene was initiated only one sixteenth as much in infected mice as in the controls. This defect was unique to the growth-hormone gene; transcription proceeded normally for other genes, such as those for thyroid-stimulating hormone and actin (an essential structural protein), in both infected and uninfected mice.

To uncover further details about the precise mechanism by which LCMV interfered with transcription of the growth-hormone gene, we needed to know more about the virus itself. What viral genes were responsible for the interference? In searching for the answer, we were helped by the fact that there are two strains of LCMV, one (Armstrong) that causes disease and one (WE) that does not. LCMV has two RNA segments, a short and a long one. Yves Rivière, Rafi Ahmed and I took the two strains and swapped short and long segments among them and injected the reassorted viruses into mice. Only the viruses that contained the short segment of the Armstrong strain caused disease.

The next logical step would be to isolate the genes that are responsible for disease. We knew that the short strand is made up of 3,700 nucleotide bases and that one end encodes the viral glycoproteins (which form the outer coats of virus particles) and the other encodes the core proteins. To find out which part is the culprit, we could excise the genes from the Armstrong strain and reinsert them into the WE strain to make recombinant viral genes. Alternatively, we could make site-specific mutations in the genes to see if we could disable the disease-causing portion. Unfortunately, neither method can yet be applied to the particular class of viruses to which LCMV belongs. Instead we have been trying to compare directly the nucleotide sequence of the short-strand RNA in the disease-causing strain with the sequence in the benign strain. Such studies have not yet revealed a discrete site to account for the disease mechanism.

We had found a persistent viral infection that caused a hormonal deficiency without visible injury to infected cells. Was this a unique phenomenon or, rather, a general one involving diverse systems, such as the nervous system, the immune system and endocrine glands other than the pituitary? We quickly found that the phenomenon was not unique. In a different strain of mice, LCMV selectively invaded the beta cells of the islets of Langerhans in the pancreas. These cells normally secrete insulin, but in infected mice insulin production was apparently impaired, and the mice developed signs of diabetes. Yet the infected islet cells appeared normal under the microscope and showed no sign of inflammation. These studies showed that a virus could persistently infect islet cells and result in a biochemical and morphological picture comparable to that of adult-onset diabetes in humans.

We obtained similar results with thyroid epithelial cells and with neurons. Linda S. Klavinskis of Scripps found that LCMV established persistent infection in thyroid epithelial cells and perturbed the production of thyroid hormone. Although the cells showed no morphological damage, the blood levels of two thyroid hormones, T_3 and T_4, were reduced, as was the amount of messenger RNA encoding thyroglobulin, a precursor of both hormones. Similarly, W. Ian Lipkin of Scripps found that LCMV infected neurons containing the neuropeptide somatostatin but not those making cholecystokinin or gammaaminobutyric acid (GABA). The infected neurons also showed no morphological injury, but they contained significantly less messenger RNA coding for somatostatin than did neurons from uninfected animals. The effect was specific to somatostatin: messenger RNA for cholecystokinin and

GABA did not decrease in infected animals. Other viruses have since been found to establish persistent infections in various organ systems, in humans as well as in animals, with analogous disruption of specialized cell functions.

In order to establish a persistent infection, viruses must be able to evade an organism's immune defenses. One common way is for the virus to infect lymphocytes and also macrophages (another type of immune cell, which digests injured cells and foreign bodies). The infection of these cells may play a role in the spread of infection within a host, in viral latency and persistence and in the transmission of viruses to uninfected individuals. In particular, we think that infection of lymphocytes may be responsible for the selective immune suppression that accompanies persistent viral infections.

Normally, when an animal has an acute LCMV infection, the viruses are attacked by virus-specific cytotoxic T cells, also known as killer T cells. The virus-specific killer T cells recognize certain viral particles that are bound to a unique "self" glycoprotein, called a major histocompatibility complex (MHC), on the surface of host cells. Each killer T cell recognizes and destroys only a specific viral antigen in combination with a specific MHC. For example, killer T cells specific to LCMV-infected cells of a specific MHC type will not kill the same MHC-type cells infected by different viruses; similarly, they will not kill cells of a different MHC type even if they are infected by LCMV. The killer T cells are primarily responsible for destroying infected cells, thereby terminating the infection.

In mice that have weak immune systems, however, the situation is markedly different. Persistent infections can be initiated by injecting LCMV into mice whose immune systems have been suppressed by such treatments as X-radiation and administration of immunosuppressive drugs. Similarly, if newborn mice are injected with the virus, their immature immune systems are unable to clear the virus, and the animals develop a lifelong persistent infection (see Figure 10.5).

When we analyzed lymphocytes from persistently infected mice, we found infectious viruses and viral genes in a small percentage of the lymphocytes. Because these viruses were able to invade lymphocytes and apparently suppress the immune response, we thought they might be different from the seed virus with which we had inoculated the animals. Ahmed and I decided to explore the question. The results of our investigation were startling: we found that the virus recovered from the lymphocytes had the distinctive ability to initiate persistent infection in adult animals. Since these animals had intact immune systems, we concluded that the virus found in the lymphocytes must be a mutant strain able to suppress selectively the immune response targeted against it. We suspect that the virus, having first initiated a persistent infection in an animal with a weak immune system, then had the chance to replicate and produce mutants that could suppress the virus-specific killer T cells.

We also made another interesting finding, one that could have important implications for the understanding of immune suppression by viruses. When we searched for LCMV in the lymphocytes of persistently infected animals, we discovered that only about 2 percent of the lymphocytes—and mainly the so-called helper T cells—contained viral genes. Of the 2 percent, only one in 50 was infectious (that is, contained replicating virus). The vast majority of LCMV-infected lymphocytes, then, harbor incomplete, nonreplicating variants of the virus. This finding suggests that a virus need knock out only a small fraction of lymphocytes in order to cause selective immune suppression. What is more, whole, replicating viruses may not be necessary; incomplete viruses may be effective in disabling lymphocytes. This behavior of LCMV bears a striking resemblance to that of the human immunodeficiency virus (HIV), the pathogen implicated in AIDS.

What mutation in the seed virus enabled some of its progeny to suppress the immune system? To find out, Mario Salvato in our laboratory sequenced the nucleic acids in the viral RNA. She established that the mutations in the long RNA segment of the virus are restricted to the section encoding the viral polymerase (the enzyme that transcribes viral DNA) or to the section encoding "Z protein" (so called because it is a type of zinc finger, a regulatory protein that binds to nucleic acid), or to both. Independently, Ahmed and his colleagues at the University of California at Los Angeles also located the immunosuppressive mutation on the large RNA segment by a different technique: reassorting the short and long segments of viral RNA from the immunosuppressive and the parental strains of LCMV.

We conclude from these experiments that variants of LCMV are spawned in lymphocytes and that they abort the production of the very killer T

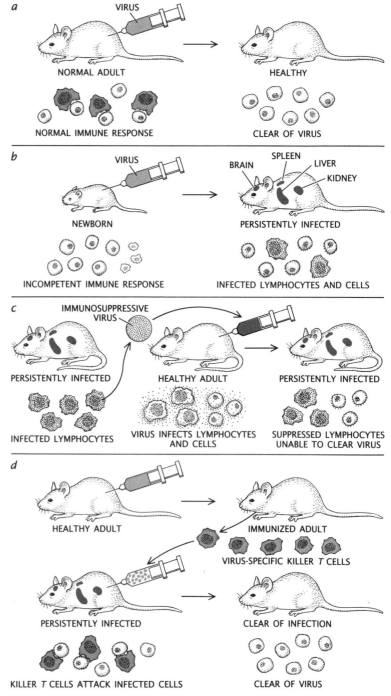

Figure 10.5 IMMUNE SUPPRESSION by LCMV. When LCMV (*blue*) was injected into normal adult mice, the animals remained healthy (*a*); infected cells were destroyed by killer *T* cells. When the virus was injected into newborn mice, the animals developed a persistent infection (*b*); the immature lymphocytes were unable to clear the virus and a subset of the lymphocytes became infected. Healthy adults injected with virus from the infected lymphocytes also became persistently infected (*c*); the virus had mutated into a strain (*red*) that infects and suppresses the lymphocytes. When uninfected, virus-specific killer *T* cells (*green*) from LCMV-immunized mice were injected into persistently infected mice, they cleared the virus (*d*).

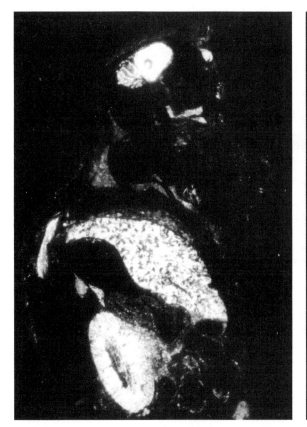 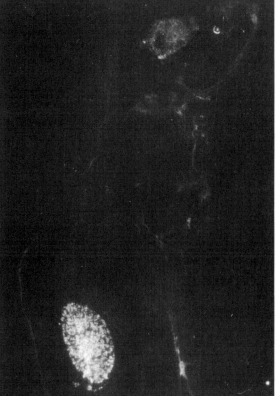

Figure 10.6 HEAVY LOAD of viral protein shows up in an autoradiograph of a longitudinal section of a mouse with a persistent LCMV infection (*left*). The virus is concentrated in the brain, liver, spleen and kidney. A second persistently infected animal was injected with killer *T*-cells from an LCMV-immunized mouse. After 21 days the animal was free of viral proteins (*right*) except in the brain, which was clear by 60 days, and the kidney, where complexes of virus bound to viral antibodies had lodged.

cells whose function it is to clear LCMV-infected cells. The variants thereby enable LCMV infection to become persistent, so that viral nucleic acids and proteins, as well as infectious viruses, accumulate in the blood and tissues. It is not yet known how other persistent or latent viruses suppress the immune response, but it seems possible that similar scenarios are played out in infections by such viruses as HIV, cytomegalovirus and hepatitis *B* virus, all of which are known to infect and alter the functions of lymphocytes.

An urgent question in the treatment of persistent infections caused by virus-induced immune suppression is whether there is any hope of repairing the damage once the immune defenses have fallen.

Our further investigations with LCMV indicate that such repair may be possible (see Figure 10.6). We injected persistently infected mice with the appropriate type of virus-specific killer *T* cells, taken from the spleen of healthy animals (of the appropriate MHC type) that had been immunized against LCMV. The results were truly remarkable: the mice became virtually free of infection. The transplanted killer *T* cells removed both the infectious virus and the deposits of viral nucleic acids. Similar cytoimmunotherapy might be applied to terminate other persistent infections; in each case, however, one must first identify the immune cell that has been disabled.

Evidence is accumulating that situations similar to

VIRUS AND HOST	TARGET CELLS	SYNDROME
LCMV IN MICE	PITUITARY CELLS	GROWTH-HORMONE DEFICIENCY, GROWTH RETARDATION, HYPOGLYCEMIA
LCMV IN MICE	HELPER *T* CELLS, KILLER *T* CELLS	SELECTIVE IMMUNE SUPPRESSION
LCMV IN MICE	THYROID FOLLICULAR CELLS	HYPOTHYROIDISM
LCMV IN MICE	BETA CELLS IN PANCREAS	DIABETES
LCMV IN MICE	NEURONS	REDUCED SOMATOSTATIN
CANINE DISTEMPER VIRUS IN MICE	UNKNOWN	OBESITY, REDUCED NOREPINEPHRINE AND DOPAMINE
MINK-CELL FOCUS-INDUCING RECOMBINANT VIRUS IN MICE	UNKNOWN	DEFORMED WHISKERS
VENEZUELAN EQUINE ENCEPHALITIS IN HAMSTERS	BETA CELLS IN PANCREAS	DIABETES
SEMLIKI FOREST VIRUS IN MICE	NEURONS	REDUCED GAMMA-AMINOBUTYRIC ACID
RABIES IN MICE	NEURONS	IRREGULAR BRAIN WAVES
COXSACKIE *B* VIRUS (OR COXSACKIELIKE SEQUENCES) IN HUMANS	MYOCARDIAL CELLS	HEART DISEASE

Figure 10.7 PROVED LINKS between persistent viral infection and specific disorders are listed. In all cases the virus was observed in vivo to interfere with the specialized, "luxury" functions of cells without disturbing their vital "housekeeping" functions.

those we have described in mice may be occurring in humans (see Figure 10.7). M. A. Preece and his colleagues at the University of London have found that some children who have had a rubella infection have retarded growth and depressed glucose metabolism. In the two boys who were studied, both abnormalities were corrected by replenishing their growth-hormone levels. Among the many afflictions of people who were exposed to rubella in utero is diabetes, which develops in nearly 20 to 30 percent of the cases. Coxsackie virus, cytomegalovi-

rus and mumps virus can acutely infect and injure the beta cells in the islets of Langerhans, and mumps virus can also attack thyroid cells; whether such viruses can persist is not known. And many common childhood viruses, such as measles, chicken pox and mumps, persist in neurons in the central nervous system throughout life.

Persistent viruses may also turn out to be implicated in damage to vital organs. In the past two years, L. Anchard and his colleagues at the Charing Cross and Westminster Medical School in London and Heinz-Peter Schultheiss, Peter H. Hofschneider and their colleagues at the University of Munich have linked Coxsackie or Coxsackielike enteroviruses to dilated heart muscle (cardiomyopathy), a lethal condition that sometimes requires a heart transplant. They found that 30 to 50 percent of a group of more than 70 patients had viral nucleic-acid sequences in their heart muscle, whereas biopsied tissue from some 40 other patients suffering from other heart-muscle diseases did not.

The evidence gathered over the past decade has demonstrated a new way in which a virus can do harm. Because we now know that viruses can disrupt the production of hormones and neurotransmitters, we think persistent viral infections may play a hitherto unsuspected role in many human diseases. Indeed, it would seem prudent to evaluate any disorder of the specialized functions of nerve cells, endocrine glands and the immune system for possible infectious causes.

The number of such disorders is vast: it includes hormonal abnormalities such as adult-onset diabetes, neuropsychiatric disorders and autoimmune diseases such as systemic lupus erythematosus and multiple sclerosis. Whether these diseases will ultimately turn out to have viral origins is impossible to say. What is clear is that one can no longer rely on the received wisdom about viral pathology to make that determination.

The Causes of Diabetes

A viral infection in a host of the appropriate genetic makeup may also trigger the autoimmune destruction of signaling cells, such as the beta cells of the endocrine pancreas, causing a loss of insulin secretion thereby leading to the development of diabetes.

. . .

Abner Louis Notkins
November, 1979

Diabetes mellitus and its complications are now thought to be the third leading cause of death in the U.S., trailing only cardiovascular disease and cancer. According to a report issued by the National Commission on Diabetes in 1976, as many as 10 million Americans, or close to 5 percent of the population, may have diabetes, and the incidence is increasing yearly. The direct and indirect effects of diabetes on the U.S. economy are enormous, exceeding $5 billion per year. If current trends continue, the average American born today will have better than one chance in five of ultimately developing the disease. The likelihood of becoming diabetic appears to double with each decade of life and with every 20 percent of excess body weight.

Moreover, although the acute and often lethal symptoms of diabetes can be controlled with insulin therapy, the long-term complications of the disease reduce life expectancy by as much as a third. Compared with nondiabetics, diabetics show a rate of blindness 25 times higher, of kidney disease 17 times higher, of gangrene five times higher and of heart disease twice as high.

Many aspects of diabetes remain mysterious, but recent work in three seemingly unrelated fields — genetics, immunology and virology — has supported the contention that diabetes is a heterogeneous group of diseases rather than a single one. This work has also indicated that diabetes arises from a complex interaction between the genetic constitution of the individual and specific environmental factors.

Diabetes mellitus is an ancient disease. The earliest description of its symptoms is found in the Ebers papyrus of Egypt, dating back to 1500 B.C. In the second century A.D. Aretaeus of Cappadocia named the disease diabetes, the Greek word meaning "to flow through a siphon." "Diabetes," he wrote, "is a strange disease that consists in the flesh and bones running together into the urine." This was an imaginative description of the striking symptoms of diabetes: a copious flow of urine accompanied by extreme thirst and hunger, but none-theless resulting in the wasting away of both muscle and fat, often ending in coma and death. In the sixth century Indian physicians recognized that the urine from diabetic patients had a sweet taste. It was not until the 18th century, however, that the sweet-tasting substance was identified as the sugar glucose and the word mellitus, or "honeyed," was added.

One of the first clues to the pathology underlying diabetes came in 1889, when Oscar Minkowski and Baron Joseph von Mering, working in Strasbourg, sought to determine whether the pancreas was es-sential to life. By careful surgical procedures they removed the pancreas from dogs. A probably apoc-ryphal story has it that the day after the operation the caretaker in the laboratory noticed that the dogs' urine attracted an unusual number of flies. In any event Minkowski and von Mering analyzed the urine and found in it high levels of glucose, indicat-ing that removal of the pancreas gave rise to a syn-drome resembling diabetes. This finding strongly implied that the pancreas was secreting a substance that regulated the metabolism of glucose.

In 1909 the hypothetical antidiabetic substance secreted by the pancreas was given a name: insulin. All attempts to alleviate experimental diabetes by feeding pancreatectomized dogs raw pancreas or by injecting them with crude pancreatic extracts, how-ever, were unsuccessful. It is now known that such experiments were bound to fail because insulin is a protein, and that when it is given orally, it is de-stroyed by protein-cleaving enzymes in the gastro-intestinal tract and by similar enzymes present in crude pancreatic extracts. As a result definitive proof of the existence of insulin was not forthcom-ing until two Canadian investigators. Frederick G. Banting and Charles H. Best, extracted insulin from dog pancreases that they had previously depleted of protein-cleaving enzymes. On July 30, 1921, Bant-ing and Best injected their pancreatic extract into a diabetic dog. Within hours the glucose level in the blood began to fall. News of the experiment spread rapidly, and in a short time insulin was being widely and successfully employed to treat the acute symptoms of diabetes mellitus in human beings. The discovery of insulin was hailed as a cure for diabetes because it lowered blood-glucose levels, controlled the acute symptoms of the disease and prevented the death from coma that sometimes came within days after the onset of symptoms.

Evidence that all was not well only became ap-parent years later. Diabetics who had been on insu-lin for a long time were found to have an unusually high incidence of heart attacks, stroke, kidney fail-ure, gangrene and blindness. Disorders of the nerves, the skin and the mouth were also common, and particularly serious complications arose during pregnancy. Insulin treatment thus controlled the early symptoms of diabetes but not the develop-ment of long-term complications.

What causes diabetes and its complications? Since the time of Banting and Best much has been learned about insulin and how it controls glu-cose levels in the blood. The pancreas (see Figure 11.1) has two distinct components: the acinar cells, which manufacture digestive enzymes and secrete them into the duodenum (the first segment of the small intestine), and the islets of Langerhans, which secrete a variety of hormones into the bloodstream. The pancreas has between one and two million islets, each islet about 200 microns in diameter and together accounting for about 2 percent of the mass of the organ. The islets are highly vascularized, and each islet cell is in close proximity to a capillary.

There are at least four different types of cells in each islet. The alpha cells, which make up about 20 percent of a typical islet, secrete the hormone gluca-gon. The beta cells, which make up about 75 per-cent of the islet, secrete insulin (see Figure 11.2). Insulin and glucagon act in different ways; whereas the secretion of insulin lowers the level of glucose in the blood, the secretion of glucagon raises it. The delta cells secrete the hormone somatostatin, which inhibits the secretion of both insulin and glucagon, and the PP cells secrete pancreatic polypeptide hor-mone, the function of which is not yet clear. Al-though the concentration of glucose in the blood is maintained at a more or less constant level by the actions of insulin, both glucagon and somatostatin play an important modulating role.

In normal individuals the concentration of glu-cose is usually less than 115 milligrams per 100 milliliters of plasma, but in diabetics it is much higher and in severe cases may reach 1,000 milli-grams. Because diabetics have a particularly diffi-cult time removing excess glucose from the blood following the ingestion of carbohydrates they can be easily diagnosed by means of a glucose-tolerance test (see Figure 11.3). A standard load of glucose is given by mouth, and the blood-glucose concentra-tion is measured periodically over the next few hours. In the normal individual the blood glucose returns to base-line levels within two hours. In the

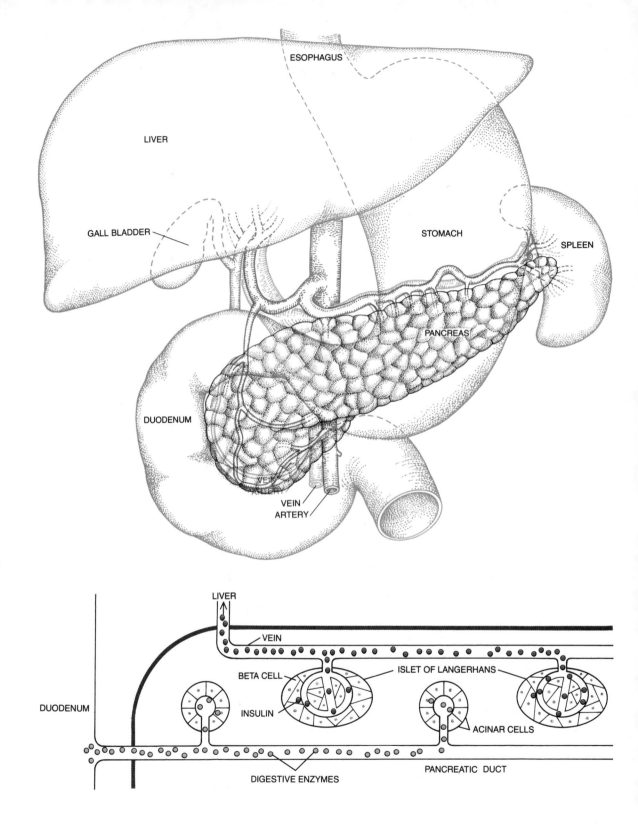

ESOPHAGUS

LIVER

GALL BLADDER

STOMACH

SPLEEN

PANCREAS

DUODENUM

VEIN
ARTERY

VEIN
ARTERY

LIVER

VEIN

BETA CELL

ISLET OF LANGERHANS

DUODENUM

INSULIN

ACINAR CELLS

DIGESTIVE ENZYMES

PANCREATIC DUCT

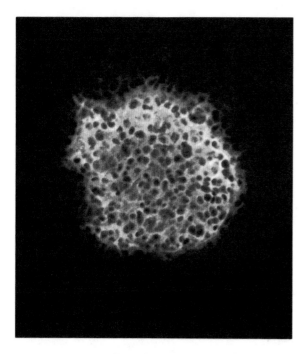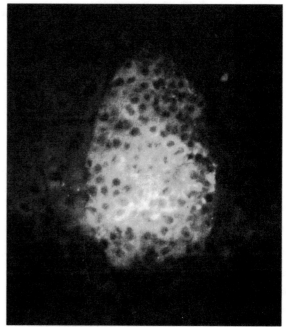

Figure 11.2 INSULIN-CONTAINING BETA CELLS within an islet of Langerhans in the pancreas glow orange in a fluorescence micrograph (*left*). The dark spot within each fluorescent cell is the cell nucleus; the surrounding acinar cells of the pancreas do not contain insulin and hence appear black. In a micrograph of a section of pancreas from a mouse infected three days earlier with encephalomyocarditis (EMC) virus (*right*) only the cells within the islet of Langerhans are infected and glow green; the surrounding acinar cells are not infected and appear dark. Such experiments have shown that viruses can destroy beta cells and induce diabetes in experimental animals.

diabetic, however, the blood-glucose values rise much higher and take longer to return to the baseline levels.

After the ingestion of a meal the beta cells of a normal individual respond to the resulting rise in blood-glucose levels by secreting more insulin into the bloodstream (see Figure 11.4). The insulin travels to the various organs of the body, interacting with specific receptors on the surface of the target cells. The binding of the hormone to its receptors initiates a series of events within the cells that results in the increased uptake of glucose into the cells, where it is converted into metabolic energy or stored as glycogen (animal starch) and fat.

It is obvious that pathological processes intervening anywhere along this pathway (such as in the pancreas, the blood or the peripheral tissues) could result in abnormal glucose metabolism. For example, the glucose in the blood would be elevated if the beta cells of the pancreas did not manufacture enough insulin, or if there were antagonists to insulin in the bloodstream, or if the peripheral tissues of the body did not respond properly to the action of insulin.

The causes of the long-term complications of diabetes are even more perplexing. One of the many complications is the thickening of the basement membrane that surrounds the wall of capillaries (see Figure 11.5). This thickening is believed to contrib-

Figure 11.1 PANCREAS is situated in the abdominal cavity just below the liver and under the stomach. It is bordered on one side by the duodenum (the first segment of the small intestine) and on the other side by the spleen. The pancreas is made up of two functionally distinct components: the acinar cells and the islets of Langerhans, as is shown in the schematic diagram. The acinar cells, which make up the bulk of the pancreas, manufacture digestive enzymes that enter the duodenum by way of the pancreatic duct. The islets of Langerhans represent only 1 to 2 percent of the total mass of the pancreas and secrete several hormones (such as somatostatin and glucagon) in addition to insulin. These hormones reach the bloodstream by way of numerous small veins in the islets of Langerhans that drain into the liver.

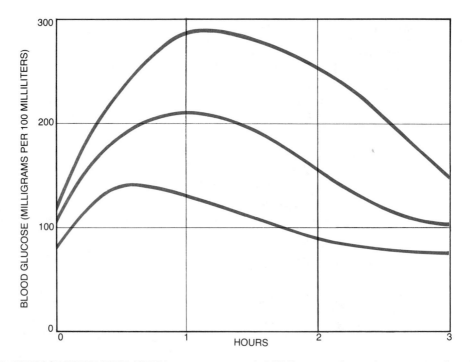

Figure 11.3 GLUCOSE-TOLERANCE TEST is a common method for diagnosing diabetes in individuals whose fasting glucose is not unequivocally elevated. Seventy-five grams of glucose are administered orally and changes in blood-glucose levels are monitored over several hours. Bottom curve represents the response of a normal individual. Middle curve shows the response of an individual who has an impaired glucose tolerance but is not considered truly diabetic. Top curve represents the response of a diabetic and shows that the blood-glucose level remains elevated.

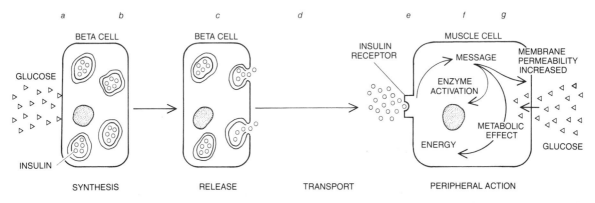

Figure 11.4 RISE IN BLOOD GLUCOSE associated with a carbohydrate meal induces beta cells in the islets of Langerhans to secrete insulin, which is carried to target cells throughout the body and binds to receptor molecules on the cell surface. This interaction triggers a series of events inside the cells that enhances the uptake of glucose from the blood and its subsequent breakdown for metabolic energy or storage as glycogen and fat. Possible defects along this pathway that could result in diabetes include destruction of beta cells (a), abnormal synthesis of insulin (b), retarded release of insulin (c), inactivation of insulin by antibodies or other blocking agents (d), altered or fewer insulin receptors (e), defective processing of the insulin message within the target cells (f) and abnormal metabolism of glucose (g).

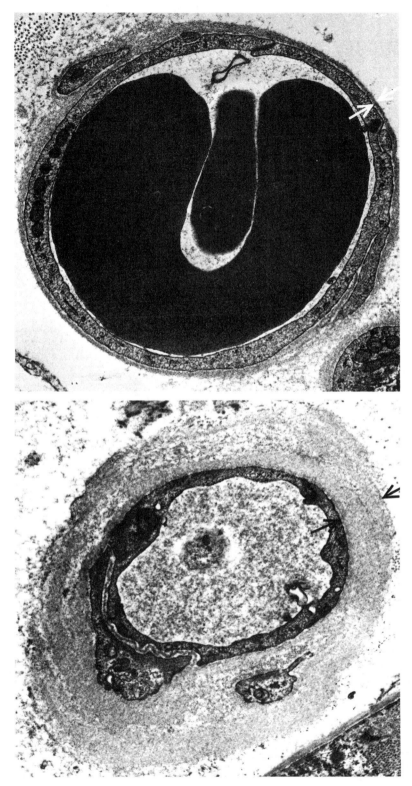

Figure 11.5 THICKENING OF BASEMENT MEMBRANE that surrounds the blood vessels is one of the long-term complications of diabetes. The micrograph at the top shows a capillary from a normal male; the basement membrane (*between arrows*) is a thin sheath surrounding the epithelial cells of the vessel. (The dark structures inside the capillary are red blood cells.) The micrograph at bottom shows a capillary from a man with diabetes of 26 years' duration and the basement membrane is markedly thickened. This change is thought to contribute to poor peripheral circulation, damage to the retina of the eye and an acceleration of the atherosclerotic process.

ute to poor peripheral circulation and to be at least partially responsible for the fact that many diabetics suffer from disorders of more than one organ. A number of hypotheses have been put forward to explain the pathological processes that give rise to complications. One hypothesis holds that in the diabetic there is a premature aging of cells, so that the diminution of function that might be expected late in life comes much earlier. Indeed, changes characteristic of aging have been observed when cells from diabetics are cultured in laboratory glassware, and it has been suggested that such changes might be genetically controlled. Another hypothesis suggests that specific metabolic intermediates of glucose, such as sorbitol, accumulate to high concentrations in tissues such as nerves and the lens of the eye. This accumulation could lead to changes in osmotic pressure, causing the cells to swell and thereby damaging the tissue.

A hypothesis that is currently receiving considerable attention contends that high concentrations of glucose in the blood cause glucose molecules to form chemical bonds with the amino groups of cellular proteins, the reaction known as glycosylation. This process was first recognized when it was found that in the blood of diabetics glycosylated forms of hemoglobin are unusually common. Many investigators are now beginning to believe proteins in other tissues, such as the eye, the nerves and the blood vessels, may become glycosylated to a much greater degree in diabetics than in nondiabetics. The critical question is whether these changes and others caused by high concentrations of glucose are genuinely responsible for the thickening of the basement membrane of the blood vessels and the other long-term complications of diabetes. The question is not merely academic; if a link between high levels of blood glucose and complications is firmly established, then the most prudent course would be strict therapeutic measures to keep glucose levels close to the normal base line.

The different and often conflicting findings about the nature of diabetes and its long-term complications have led many to conclude that it is not a single disease but rather a heterogeneous group of diseases, all of which ultimately lead to an elevation of glucose in the blood. Indeed, two major types of diabetes have been distinguished on the basis of clinical evidence: the maturity-onset type and the juvenile-onset type. Long-term complications can develop in both types, although there is

considerable variation among individuals (see Figure 11.6).

Maturity-onset diabetes is the much more prevalent type, representing more than 90 percent of all the cases. It most often occurs in people who are over 40 and overweight The onset is slow, and pathological changes in the pancreas are not always apparent. Moreover, the clinical manifestations of maturity-onset diabetes are often mild, and the high glucose levels in the blood can usually be controlled by diet alone. Although diabetes is traditionally viewed as a condition caused by a deficiency of insulin, many maturity-onset diabetics have a sufficiency or even a surplus of the hormone in the blood. For these individuals diabetes arises not from a shortage of insulin but probably from defects in the molecular machinery that mediates the action of insulin on its target cells. Maturity-onset diabetes is accordingly referred to as non-insulin-dependent diabetes.

Again a number of hypotheses have been put forward to explain maturity-onset diabetes, and there is little unanimity in the field. One of the most intriguing hypotheses comes from the work of Jesse Roth, C. Ronald Kahn and their colleagues at the National Institute of Arthritis, Metabolism, and Digestive Diseases. By measuring the binding to target cells of radioactively labeled insulin molecules they showed that the number of insulin receptors was decreased in obese patients with maturity-onset diabetes. When the patients were put on a weight-reducing diet, however, the number of insulin receptors returned to normal. Roth and Kahn contend that the increased food intake associated with obesity leads initially to the secretion of an excessive amount of insulin into the circulation. The secreted insulin then acts through a negative-feedback process of some kind to reduce the number of insulin receptors on the target cells. This decrease in receptors is thought to make the cells less responsive to insulin and hence less capable of utilizing glucose. Other investigators propose, however, that the primary defect actually arises within the target cells after insulin has bound to the receptors. Regardless of the mechanism, most would agree that the proper control of diet and body weight is of major importance in treating this form of diabetes.

Juvenile-onset diabetes is much less common than the maturity-onset type, representing well under 10 percent of all the cases. It usually develops in people younger than 20, and its onset is more abrupt. The disease is characterized by a marked

FEATURES	JUVENILE-ONSET (INSULIN-DEPENDENT)	MATURITY-ONSET (NON-INSULIN-DEPENDENT)
AGE AT ONSET	USUALLY UNDER 20	USUALLY OVER 40
PROPORTION OF ALL DIABETICS	LESS THAN 10 PERCENT	GREATER THAN 90 PERCENT
SEASONAL TREND	FALL AND WINTER	NONE
APPEARANCE OF SYMPTOMS	ACUTE OR SUBACUTE	SLOW
METABOLIC KETOACIDOSIS	FREQUENT	RARE
OBESITY AT ONSET	UNCOMMON	COMMON
BETA CELLS	DECREASED	VARIABLE
INSULIN	DECREASED	VARIABLE
INFLAMMATORY CELLS IN ISLETS	PRESENT INITIALLY	ABSENT
FAMILY HISTORY OF DIABETES	UNCOMMON	COMMON
HLA ASSOCIATION	YES	NO
ANTIBODY TO ISLET CELLS	YES	NO

Figure 11.6 DIFFERENCES between juvenile-onset and maturity-onset diabetes are summarized. The maturity- onset form is commoner and not as severe. It does not require injections of insulin.

decline in the number of beta cells in the pancreas (often to less than 10 percent of normal), leading to a deficiency of insulin and an elevation of glucose in the blood. The deficiency of insulin accelerates the breakdown of the body's reserve of fat, resulting in the production of ketones and organic acids. These metabolites lower the pH of the blood, producing a condition known as diabetic ketoacidosis that can result in death. Because injections of insulin are required to regulate the level of blood glucose, this form of diabetes, which is generally more severe, is referred to as insulin-dependent diabetes.

Although there is no general agreement on the cause of juvenile-onset diabetes, the reduction in the number of beta cells and the decline in insulin levels long ago suggested that the primary deficit is at the level of the beta cell. Over the past few years new information has emerged on some of the factors that might cause beta-cell damage. The new leads concerning the relation between genetic and environmental factors are what I shall now discuss in depth.

It has been known for some time that maturity-onset diabetes tends to run in families and that the likelihood of an individual's developing the disease is increased if one parent is diabetic and further increased if both parents are. In contrast, the likeli-

hood of developing juvenile-onset diabetes is not substantially increased in the offspring of diabetic parents. The precise risk of developing the different forms of diabetes has been difficult to establish, however, because of the variety of the diagnostic criteria and because such variables as diet, obesity, age, sex and ethnic background are not always taken into account. Moreover, since family members share the same diet and environment, a high incidence of diabetes in a given family does not necessarily prove that genetic factors are involved.

In the hope of distinguishing between genetic and environmental factors in diabetes geneticists began some 40 years ago to study identical twins: offspring derived from the same egg who share the same genes. If diabetes were caused solely by inheritance, then if one identical twin developed the disease, the other twin would be expected to develop it. The degree of genetic involvement in diabetes can therefore be estimated from the degree of concordance (both twins developing diabetes) as opposed to discordance (only one twin developing diabetes).

In the early 1970's David A. Pyke and his colleagues at King's College Hospital in London reported their findings on more than 100 pairs of identical twins, the largest series of twins ever studied. They found that when one twin of a pair developed diabetes after age 50, the other twin developed the disease within several years in almost every case. If one twin developed the disease before age 40, however, the other twin developed it in only half of the cases. Strikingly, the majority of the twins over 50 had non-insulin-dependent diabetes, whereas most of those under 40 had insulin-dependent diabetes. Pyke's findings generated considerable excitement because they demonstrated that genetic factors are predominant in maturity-onset diabetes but that additional factors, presumably environmental, are needed to trigger juvenile-onset diabetes.

A different approach to the genetics of diabetes has been to investigate the relation between diabetes and the histocompatibility antigens. These antigens—proteins on the surface of all body cells with nuclei—are responsible for the fact that tissue transplanted from one individual to an unrelated individual will be recognized as foreign and will be rejected by the recipient's immune system (see Figure 11.7). In man the histocompatibility antigens are referred to as the HLA system. The genes coding for the HLA antigens are on chromosome No. 6 and occupy four loci along the chromosome designated *A, B, C* and *D*. The genes at a given locus are not always the same, and the different forms of each gene are referred to as alleles. An individual can have two different alleles at a given locus, one allele contributed by each parent. In the HLA system both alleles are expressed as cell-surface proteins, which can be identified by laboratory tests and can serve as the basis for the typing of tissues for transplantation.

A few years after tissue typing became a common procedure a completely unexpected discovery was made. Certain of the HLA antigens were found at unusually high frequency in patients with specific diseases. For example, the risk of developing the deformity of the spine known as ankylosing spondylitis was found to be 100 times greater in individuals carrying the HLA antigen *B27*. This observation spurred investigators in many parts of the world to look for an association between HLA antigens and other diseases, including diabetes. Jørn Nerup and his associates at the Steno Memorial Hospital in Copenhagen found that the HLA antigens *B8* and *B15* were two to three times commoner in diabetics than they were in nondiabetics. When the Danish workers further analyzed their data, they made an interesting discovery. The increased frequency of these HLA antigens was associated only with juvenile-onset diabetes; there was no change in the frequency of the antigens associated with the maturity-onset disease. It soon became apparent that there was an even stronger association between juvenile-onset diabetes and HLA antigens at the *D*

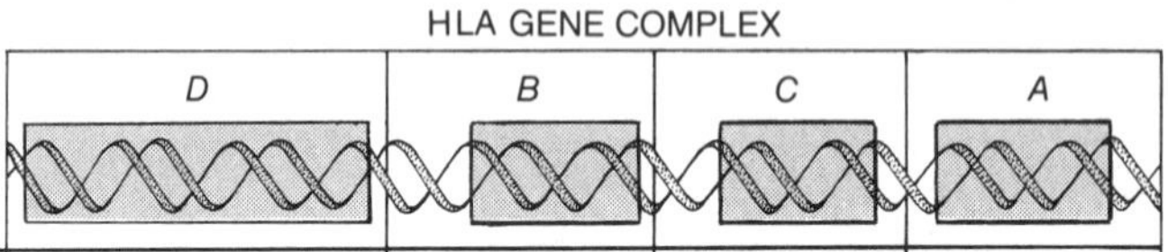

HLA GENE COMPLEX

	D	B	C	A
ALLELES ASSOCIATED WITH INCREASED SUSCEPTIBILITY	DW 3 DW 4 DRW 3 DRW 4	B 8 B 15 B 18 B 40 BW 22	CW 3	A 1 A 2
ALLELES ASSOCIATED WITH DECREASED SUSCEPTIBILITY	DW 2 DRW 2	B 5 B 7		A 11

Figure 11.7 HISTOCOMPATIBILITY ANTIGENS are cell-surface proteins that provide each individual's tissues with a unique biological label. The major category of these antigens in man, known as the HLA complex, are coded for by a series of genes on chromosome No. 6. The genes at each position or locus on the two homologous chromosomes are not always identical, and a large number of alternate genes (alleles) may be present in a population. Certain HLA alleles have been shown to be associated with an increased or decreased risk of developing juvenile-onset diabetes. Such correlations imply that genes closely linked to the high-risk HLA alleles (perhaps those genes controlling the immune system) may play a role in the genesis of diabetes.

locus. Moreover, when more than one high-risk allele was present in the same individual, the likelihood of developing juvenile-onset diabetes was increased by as much as tenfold.

Further support for an association between the HLA system and diabetes has come from the work of A. G. Cudworth of St. Bartholomew's Hospital in London, Jose Barbosa of the University of Minnesota Medical School and Pablo Rubinstein of the New York Blood Center, all of whom looked at families in which two or more siblings had juvenile-onset diabetes. When they compared diabetic and nondiabetic siblings in the same family, they found that the siblings with diabetes had inherited identical groups of HLA alleles in a significantly greater percentage of cases than the siblings without diabetes. Taken together with the fact that several different high-risk alleles have been identified, it now appears that one or more genes in close proximity to the HLA complex along chromosome No. 6 actually may be the important determinants of juvenile-onset diabetes. Recently, however, the situation has been complicated by the finding that certain HLA alleles are associated with a significant decline in the incidence of juvenile-onset diabetes, implying protective genes exist as well.

How genes might act to cause diabetes or prevent it is far from clear. One clue comes from the work of Hugh O. McDevitt of the Stanford University School of Medicine and Baruj Benacerraf of the Harvard Medical School. They found that specific genes in the mouse control the immune-response (Ir) genes and that these genes are in the same region of the chromosome as the genes for the major histocompatibility antigens of the mouse. On the basis of this evidence it is conceivable that genetically controlled differences in the immune response influence the development of diabetes. For example, the high-risk alleles associated with the HLA complex might code for a deficient immune response to agents that preferentially attack beta cells, thereby allowing beta-cell damage and diabetes to result. Conversely, the protective alleles might enhance the host's immune response to such invaders.

There are other ways the immune response might influence the development of diabetes. Under certain circumstances the host can turn against itself and react immunologically to its own proteins and so damage its own tissues; the phenomenon is known as autoimmunity. Some five years ago G. Franco Bottazzo of the Middlesex Hospital Medical School in London, Richard Lendrum of St. Mary's Hospital in London and James C. Irvine of the Royal Infirmary in Edinburgh discovered that serum from newly diagnosed juvenile-onset diabetics contains an antibody that reacts with the alpha, beta and delta cells in the islets of Langerhans from normal, nondiabetic people. The fraction of juvenile-onset diabetics who possess this autoantibody is as high as 85 percent at the time of diagnosis but decreases to less than 25 percent after two years. Patients with maturity-onset diabetes, however, rarely possess the islet-cell antibody.

It is tempting to speculate that the high-risk HLA alleles associated with juvenile-onset diabetes influence the development of the islet-cell antibody. What, however, triggers the production of the antibody and what role does it play in causing diabetes? Some investigators believe an imbalance among the cells of the immune system is responsible for the development of islet-cell antibody. Others contend that the antibody represents an immune response to components of the islet cells that have been altered by viruses or toxic chemicals. Still others argue that the islet-cell antibody has little to do with the basic disease process, since it is present in the serum of some individuals who do not show any sign of diabetes.

The controversy may be resolved by some forthcoming experiments. Ake Lernmark and his colleagues at the University of Chicago recently demonstrated that the islet-cell antibody can react with antigens on the surface of beta cells grown in culture. These workers plan to exploit the cell-culture system to determine whether the islet-cell antibody can actually destroy beta cells. Other investigators are pursuing the possibility that the critical role in beta-cell damage is played not by antibody but by white blood cells such as lymphocytes and macrophages. Regardless of whether autoimmunity is an important cause of beta-cell damage, the islet-cell antibody appears to be a valuable marker for identifying juvenile-onset diabetes and differentiating it from maturity-onset diabetes.

Both the genetic and the immunological studies I have mentioned raised the possibility that an environmental agent such as a virus might trigger juvenile-onset diabetes, perhaps by inducing an autoimmune reaction. In 1965 Willy Gepts of the University of Brussels examined the pancreas of a number of juvenile-onset diabetics who had died shortly after the appearance of their disease. Gepts found in the islets of Langerhans of many of the patients white blood cells typical of what one might expect in response to infection or an autoimmune reaction. Additional support for an infectious process in diabetes comes from the observation that juvenile-onset diabetes often begins abruptly and that the incidence of the disease is higher in fall and winter, when infectious diseases are more prevalent.

The possibility that the infectious agent might be a virus goes back to the turn of the century, when H. F. Harris, a Philadelphia physician, observed that one of his patients developed diabetes shortly after having had mumps. Since that time there have been scattered reports documenting a temporal relation between the onset of certain viral infections and the subsequent development of diabetes. The virus most often implicated has been that of mumps, but there is still no firm evidence that the relation between mumps and diabetes is anything more than a chance association. Indeed, if the mumps virus does infect beta cells and cause diabetes in human beings, it must do so under very special circumstances. In such cases a rare strain of mumps virus may be involved, or the individual who develops diabetes may have an unusual and possibly genetically determined susceptibility to the virus.

Evidence that another common virus, that of rubella ("German measles"), might be a rare cause of diabetes comes from the more recent work of Margaret Menser, Jill Forrest and their colleagues in Australia. They found that in children and adults who had contracted rubella while still in the uterus the incidence of diabetes was considerably higher than normal. Because congenital rubella is notorious for giving rise to a variety of malformations, however, the cause of diabetes in these individuals is far from clear.

Probably the best evidence that viruses may be involved in the genesis of diabetes has come from studies with experimental animals. In 1968 John E. Craighead, who is now at the University of Vermont College of Medicine, was studying a variant of encephalomyocarditis (EMC) virus. This virus causes encephalitis and myocarditis in mice and occasional febrile illness in human beings. Craighead found that many of the mice infected with the EMC variant developed diabetes. When he examined the pancreas of the infected animals, he noticed inflammatory white cells in the islets of Langerhans and

found that many of the beta cells were damaged. Proof that the virus was actually responsible for these changes in the islets came from experiments conducted by Kozaburo Hayashi, a visiting fellow in my laboratory at the National Institutes of Health. Hayashi prepared an antibody to EMC virus and labeled it with fluorescein dye, which fluoresces a brilliant green when it is irradiated with ultraviolet (see Figure 11.8). He removed the pancreas from the infected animals and incubated sections of the organ with the labeled antibody. The antibody bound only to cells containing EMC virus, so that he was able to identify the infected cells by their fluorescence. His experiments showed that EMC virus had indeed infected beta cells and that the replication of the virus in these cells was primarily responsible for their destruction.

Subsequent work in Craighead's laboratory and mine revealed that the initial effect of the virus-induced destruction of beta cells is the release of large amounts of insulin into the circulation, which lowers the level of glucose in the blood. Within a few days, however, in animal's insulin reserve is depleted, in some cases to less than 10 percent of normal, and the glucose level rises. Many of the animals now begin to excrete glucose in their urine and to consume increasing amounts of water and food, the symptoms of juvenile-onset diabetes.

The investigation of the involvement of viruses in the genesis of diabetes took a new turn when it became apparent that only certain inbred strains of mice develop the disease. Three postdoctoral fellows in my laboratory, Wark Boucher, Michael Ross and Takashi Onodera, investigated this phenomenon in detail. They infected 24 different inbred strains of mice with EMC virus and divided the strains into three groups on the basis of their glucose response. The first group, designated "susceptible," responded to the infection by manifesting elevated levels of glucose in the blood. the second group, designated "glucose-intolerant," showed abnormally high glucose levels after infection only when a large glucose load was administered. The third group, designated "resistant," showed no sign of diabetes after infection.

Breeding experiments revealed that the differences among strains of mice in susceptibility to EMC-induced diabetes are genetically controlled (see Figure 11.9). When mice from two susceptible strains were mated, their offspring were also susceptible. Similarly, the offspring of two resistant

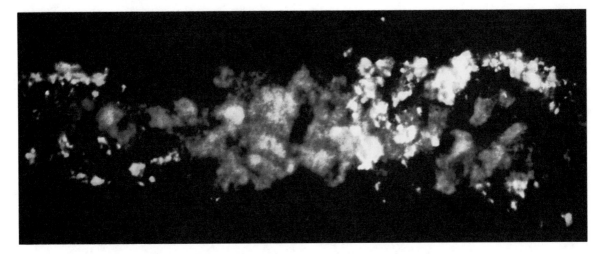

Figure 11.8 VIRUS-INFECTED BETA CELLS are revealed in this immunofluorescence micrograph. A section of pancreas from a mouse infected with reovirus was stained with two antibodies: a rhodamine-labeled antibody to insulin and a fluorescein-labeled antibody to the virus. Depending on the filter employed in the fluorescence microscope, the beta cells could be made to glow orange and the virus-infected cells to glow green. In the double exposure shown here the fluorescent colors mix so that some of the virus-infected beta cells appear yellow.

strains were resistant. A cross between a susceptible mouse and a resistant one yielded a first generation that was resistant to the development of virus-induced diabetes. When the first-generation mice were mated with one another, however, many of their offspring were susceptible, indicating that susceptibility to virus-induced diabetes had been inherited as a recessive trait. A backcross experiment was also performed: when the resistant first-generation mice were crossed with the resistant parental strain, none of the offspring developed diabetes. When the resistant first-generation mice were crossed with the susceptible parent strain, however, approximately half of the offspring developed diabetes, suggesting that a single gene locus plays a major role in controlling susceptibility to EMC-induced diabetes.

As often happens in science, these studies answered some questions and raised many others. Ji-Won Yoon, a staff scientist in my laboratory, and Onodera wanted to know how inheritance actually influences which mice develop diabetes. One possibility was that certain genes control the susceptibility of beta cells to EMC infection. The more susceptible the cells are, the more of them would be infected and the greater would be the likelihood that the mouse would develop diabetes. To see if

this was the case, sections of pancreas from infected mice were stained with fluorescent antibody to EMC virus. Counts of infected cells showed that islets from susceptible strains of mice (such as *SJR*) contained approximately 10 times more infected beta cells than islets from diabetes-resistant strains (such as *C57B:*).

These experiments did not prove, however, that differences among strains of mice in susceptibility to the virus resided at the level of the beta cell; it was possible that the differences were at the level of the immune response. In order to differentiate between these alternatives we grew beta cells from different strains of mice in tissue culture and then examined their susceptibility to infection. In this way we were able to circumvent the immune response of the mice. When cultures highly enriched in beta cells were infected with EMC virus, we found that the beta cells from susceptible strains were destroyed more readily than those from resistant strains.

A clue to what makes the beta cells from different strains of mice differentially susceptible to EMC infection has been provided by recent work in my laboratory. In 1959 John J. Holland, who is now at the University of California at San Diego, showed that viruses would not attach to and infect certain

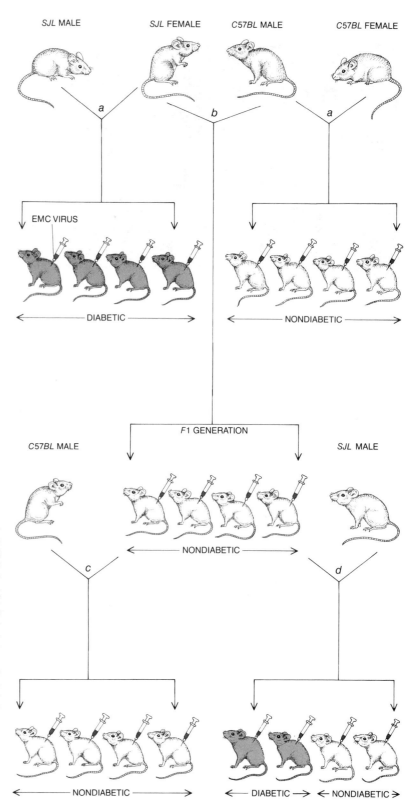

Figure 11.9 GENETIC SUSCEPTI-
BILITY to virus-induced diabetes
can be demonstrated by breeding
experiments in mice. Mice of the
strain designated *SJL* develop diabetes when they are infected with
the EMC virus, whereas those of
the strain designated *C57BL* do not
(*a*). When the susceptible and resistant strains are crossed, their offspring (the *F*1 generation) are resistant to EMC-induced diabetes (*b*).
Similarly, a backcross between the
*F*1 mice and the resistant parental
strain gives rise to resistant offspring (*c*). When the *F*1 mice are
backcrossed with the susceptible
parental strain, however, a high
percentage of the offspring are susceptible (*d*). These results suggestion that susceptibility to EMC-induced diabetes is inherited as a
recessive trait.

cells unless those cells possessed viral "receptors": cell-surface proteins of some kind that the virus recognizes and binds to before infecting the cell. Recently Ruben Chairez, working as a visiting scientist in my laboratory, found that two to three times more EMC virus bound to beta cells from diabetes-susceptible mice than bound to beta cells from diabetes-resistant mice. This finding suggests that the degree of susceptibility to the virus might be a function of the number or type of viral receptors on the surface of the beta cells. These experiments must be interpreted with caution: it is still not possible to obtain pure cultures of beta cells, and the number of viral receptors may change when the cells are grown in culture. Nevertheless, the viral-receptor experiments are a beginning and, if they are confirmed, they will provide a plausible explanation of why only animals with the "right" genetic makeup develop virus-induced diabetes. It is certainly intriguing to speculate that similar factors might operate in human beings, and that the high-risk alleles associated with the HLA complex might control the susceptibility of human beta cells to viral infection.

With the information gained from the EMC model we began to look for other viruses that might induce diabetes in mice, particularly those viruses that also infect human beings. One likely candidate was the Coxsackie family of viruses. First isolated in the 1940's from a group of patients living in Coxsackie. N.Y., these viruses have RNA as their genetic material. They can cause upper-respiratory distress and muscle pain and can infect the heart and the brain. Soon after the discovery of the Coxsackie viruses Gilbert Dalldorf of the New York State Department of Health and A. M. Pappenheimer of the Harvard Medical School found that infection of mice with members of the virus group designated Coxsackie B caused severe destruction of the acinar cells in the pancreas but spared the adjoining islets of Langerhans. Twenty years later D. Robert Gamble of the West Park Hospital at Epsom in England, who had thought for some time that there might be a connection between Coxsackie virus infection in children and diabetes, and George E. Burch of the Tulane University School of Medicine undertook to repeat those experiments. They too found destruction of acinar cells, and they also noted that some cells in the islets of Langerhans were damaged. Gamble observed that there was a transient rise in the levels of glucose in the blood of some of the infected mice, but he had difficulty reproducing his experimental results.

Attempts in my laboratory to induce diabetes in mice with Coxsackie B virus also failed. Then Yoon, Onodera and I decided to take a different approach: to make the virus more virulent by adapting it to the beta cells. Exploiting the fact that beta cells can be grown in culture, we infected the cultures with a strain of Coxsackie virus known as B4. After several days we harvested the virus and infected fresh beta-cell cultures with it. We repeated the process several times. Initially the virus replicated poorly, but with repeated passaging through the cultures it became more virulent, presumably because the passaging selected for variants of the virus that replicated better in beta cells (see Figure 11.10).

When the virus had been passaged 14 times, it was injected into mice. Within a week more than half of the animals had developed diabetes: they had low insulin levels and high glucose levels, many of the islets showed damage and viral antigens were detected in some of the beta cells. It soon became apparent that our success in inducing diabetes was due not only to the adaptation of Coxsackie B4 virus but also to the selection of the appropriate strain of mouse. As in the case of EMC virus, only certain strains of mice developed diabetes when they were infected with Coxsackie B4. Once again genetic factors were clearly influencing susceptibility to diabetes.

Another virus that we tested for its ability to induce diabetes was reovirus. The prefix reo- stands for respiratory-entero-orphan virus. The virus was so named in 1959 by Albert B. Sabin of the University of Cincinnati College of Medicine because it was found in the respiratory and gastrointestinal tract of many children and yet appeared to be an "orphan": it did not cause serious illness. In mice, however, reovirus attacks the acinar cells of the pancreas, although it spares the islets of Langerhans. Encouraged by our success in making Coxsackie B4 virus more virulent for beta cells by passaging it several times through beta-cell cultures, we did the same thing with reovirus before injecting it into mice. When we examined the pancreas of the reovirus-infected mice several days later, we found virus particles not only in beta cells (see Figure 11.11), but also in alpha and delta cells. The total number of islet cells damaged by reovirus was considerably less than that damaged by Coxsackie B4 infection, and the blood-glucose levels of most of the mice were relatively normal. When the reovirus-infected mice were stressed by injection with a large dose of glucose, however, it became quite clear that their ability to metabolize glucose had become

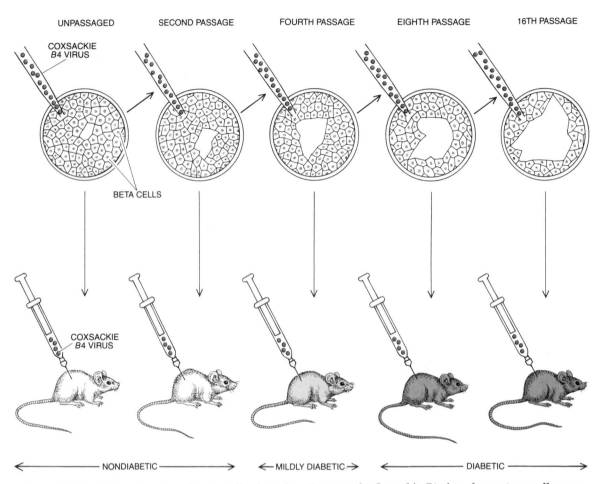

UNPASSAGED   SECOND PASSAGE   FOURTH PASSAGE   EIGHTH PASSAGE   16TH PASSAGE

COXSACKIE
B4 VIRUS

BETA CELLS

COXSACKIE
B4 VIRUS

←————— NONDIABETIC —————→   ←— MILDLY DIABETIC —→   ←————— DIABETIC —————→

Figure 11.10 PASSAGING A VIRUS repeatedly through cultures of beta cells increases its ability to induce diabetes. It is thought that passaging selects for those variants of the virus that reproduce most successfully in beta cells. For example, Coxsackie *B*4 virus does not normally cause diabetes in mice. If the virus is passaged several times, however, its capacity to kill beta cells and cause diabetes is increased.

impaired. The glucose levels remained elevated for a considerably longer period of time than normal, a condition resembling the abnormal glucose tolerance of human diabetics.

Elliot J. Rayfield of the Mount Sinai School of Medicine in New York observed similar subtle shifts in the metabolism of glucose and the release of insulin in hamsters infected with Venezuelan equine encephalitis virus, which occasionally causes fever, headache and damage to the nervous system in human beings. Taken together these studies show that variants of several common viruses known to infect human beings can attack islet cells

in experimental animals and induce symptoms resembling those of diabetes.

The first direct evidence that viruses are capable of causing diabetes in human beings came in the spring of 1978. A 10-year-old boy was brought to a hospital in the Washington, D.C., area with an illness lasting for three days that resembled influenza. He was admitted because he suffered from severe lethargy bordering on coma. Laboratory tests revealed that the child had very high levels of blood glucose (600 milligrams per 100 milliliters of plasma) and was suffering from diabetic ketoacid-

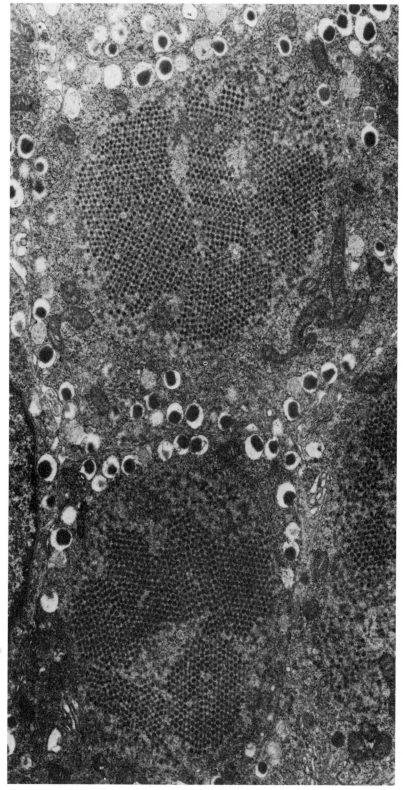

Figure 11.11 REOVIRUS PARTI-CLES form large crystalline arrays inside a beta cell from a virus-infected mouse, as is shown in this micrograph. Also visible in the cytoplasm of the cell are numerous secretory vacuoles containing dark granules of insulin. Although reovirus infection does not induce severe diabetes in mice, it does impair the animals' ability to remove glucose from the blood as measured by glucose-tolerance tests. The magnification is 17,300 diameters.

osis. In spite of intensive therapy his condition rapidly deteriorated, and within a week he died. At autopsy Robert Marshall Austin of the National Naval Medical Center found inflammatory white cells in the Langerhans and observed that many of the beta cells had been destroyed.

Since these signs were reminiscent of those observed in mice infected with diabetes-causing viruses, Austin froze a small piece of the child's pancreas for further examination. A few weeks later he told us about the case, and we eagerly agreed to look for a virus (see Figure 11.12). We homogenized the sample of pancreas and added the homogenate to tissue-culture cells known to be sensitive to a variety of viruses. Within a few days the cells showed the signs characteristic of infection. With standard techniques of virology we then succeeded in isolating from the cells a virus that had properties similar to but not identical with the virus Coxsackie B4.

In order to prove that the virus actually came from the patient and was not an inadvertent laboratory contaminant we looked for signs of infection in the samples of the patient's blood that had been taken during the course of the disease. It is well known that early in the course of a viral infection there is little if any antibody to the virus in the patient's serum because the immune system has not had enough time to respond to the invader. Later on specific antibody can easily be detected, and a rise in the levels of antibody over a period of several weeks usually implicates the virus as the cause of the illness. Serum obtained from the diabetic child on his admission to the hospital contained no antibody to Coxsackie virus, but there were significant amounts of antibody in serum obtained about a week later. This finding strongly implied that the virus was not a laboratory contaminant, but it left open the possibility that the child's infection might have been fortuitous and unrelated to the cause of his diabetes.

Here our animal model proved to be invaluable. Strains of mice known to be either susceptible or resistant to diabetes induced by Coxsackie B4 were inoculated with the virus isolated from the child's pancreas. Within a week a high percentage of the susceptible mice had developed diabetes but the resistant mice had not. In the susceptible mice inflammatory white cells were seen in the islets of Langerhans and viral antigens were detected in the

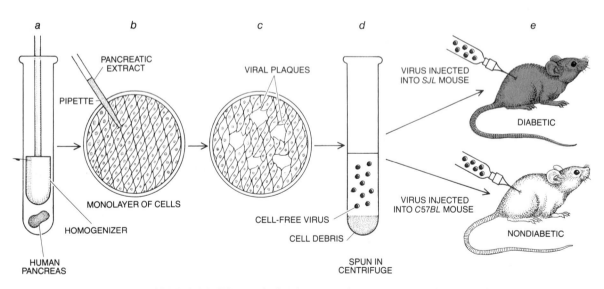

Figure 11.12 HUMAN DIABETES VIRUS was isolated from the pancreas of a boy who had died of juvenile-onset diabetes. A small piece of the pancreas was homogenized (a) and the extract placed on a layer of tissue-culture cells sensitive to a variety of viruses (b). Within several days the dissolution of virus-infected cells gave rise to plaques in the sheet of cells, indicating that a virus was present (c). The virus was separated by spinning in a centrifuge (d) and injected into mice known to be susceptible or resistant to Coxsackie B4-induced diabetes. Only the susceptible mice developed diabetes (e), implying that the child's diabetes had been induced by the virus isolated from his pancreas.

beta cells. These experiments gave strong evidence that the child's diabetes had indeed been induced by the virus.

Nevertheless, diabetes does not seem to be a common consequence of Coxsackie B4 infection. Almost 50 percent of the adult U.S. population has at one time or another been exposed to the virus, yet most people come through such infections with no apparent damage to the pancreas and no diabetes. Moreover, many juvenile-onset diabetics do not have antibody to Coxsackie B4. The case I have described might therefore have been a rare exception in which there was the right combination of a virus toxic to beta cells and a genetically susceptible individual. On the other hand, viruses capable of killing beta cells might be far commoner than has been suspected, giving rise to a spectrum of disease ranging from frequent subclinical infections with minimal beta-cell damage to relatively rare instances of moderate to severe beta-cell damage and overt diabetes. The situation might be analogous to the effects of the virus of poliomyelitis before the development of vaccines against it: many people had subclinical infections but only a few developed paralytic disease. There is, however, no evidence that a healthy individual can "catch" diabetes from an already diabetic person.

Since many juvenile-onset diabetics do not have antibody to Coxsackie B4, the search is on for other viruses that cause diabetes in man. It may be that a series of viral infections in childhood, each infection causing some beta-cell damage, finally results in overt diabetes when the reserve of beta cells has been sufficiently depleted. Along these lines it is interesting to speculate whether some children are born with a relative deficiency of beta cells or an impaired capacity to repair or regenerate beta cells once they are damaged. If that is the case, viruses might more readily produce diabetes in these already deficient individuals.

It is also possible that viruses may be only one of the many causes of diabetes (and perhaps a minor one) and that other insults from the environment such as drugs and toxic chemicals might similarly damage beta cells and give rise to diabetes. In this connection it has been known for 35 years that the drug alloxan can destroy beta cells and induce diabetes in experimental animals. The drug is highly selective in its effects: damage to beta cells can be observed within minutes after injection. In the early 1960's another drug, streptozotocin, was also shown to be toxic to beta cells and capable of inducing diabetes in experimental animals. Recently Arthur A. Like and Aldo A. Rossini of the University of Massachusetts Medical School showed that whereas a single large dose of streptozotocin is directly toxic to the beta cells of experimental animals, multiple small doses appear to act indirectly, presumably by altering the beta cells so that they become vulnerable to attack by the animal's own immune system. Under the latter conditions only certain inbred strains of mice show signs of inflammation in their islets and develop diabetes, again suggesting the importance of genetic factors.

Paradoxically, because of its toxicity and its high specificity for beta cells, streptozotocin has been used with some success to treat a rare tumor of human beta cells known as an insulinoma. In general, however, drugs and chemicals that kill beta cells were employed merely as tools for investigating the pathogenesis of diabetes in experimental animals. Then in 1975 a rodent poison known as Vacor, which has a molecular structure resembling that of streptozotocin, was introduced into the U.S. Since that time it has been accidentally or deliberately ingested by a number of people with serious consequences. Some died, and as many as 20 of the survivors developed acute diabetes calling for treatment with insulin. In two of the fatal cases examined at autopsy there was clear evidence of beta-cell destruction. A small number of other drugs and chemicals have been shown to be toxic to beta cells, although in general the damage is mild and transient.

What of the thousands of other chemicals, natural and man-made, to which human beings are exposed daily and whose effects on the beta cells have not been examined? Do some of them damage the beta cells and cause juvenile-onset diabetes? If they do, the causes of diabetes could turn out to be as numerous as those of the common cold. In any case it seems unlikely that anything as simple as a vaccine to prevent diabetes will be available in the near future.

To sum up, work in laboratories around the world has now firmly established that diabetes is not a simple disease with a single etiology. Even the juvenile-onset type may have multiple causes. Of major importance are the recent findings that genes closely linked to the HLA complex influence the risk of developing juvenile-onset diabetes and that most newly diagnosed patients have in their serum an autoantibody to islet cells. The current hope is that it will eventually be possible to identify those individuals who are most susceptible to beta-cell damage and find some way to protect them.

A second important implication is that although some cases of juvenile-onset diabetes may be explained primarily on a genetic basis and others on an environmental basis, still other cases appear to arise from a complex interaction between the genetic background of the individual and his environ-ment. The relative importance of different environmental insults, such as viruses and toxic chemicals, and of genetic factors and autoimmunity is still not clear, but research in these areas is receiving increased attention. Although it may be some time before it is possible to prevent or cure juvenile-onset diabetes and its complications, the mysteries that have long surrounded this ancient disease are gradually being dispelled.

Glucose and Aging

Glucose in high concentrations is toxic to cells and tissues. One aspect of this toxicity is the ability of glucose to form covalent links with proteins, nonenzymatically, and so alter protein structure and function. Such changes may be important in the aging process.

• • •

Anthony Cerami, Helen Vlassara and Michael Brownlee
May, 1987

As people age, their cells and tissues change in ways that lead to the body's decline and death. The cells become less efficient and less able to replace damaged materials. At the same time the tissues stiffen. For example, the lungs and the heart muscle expand less successfully, the blood vessels become increasingly rigid and the ligaments and tendons tighten. Older people are also more likely to develop cataracts, atherosclerosis and cancer, among other disorders.

Few investigators would attribute such diverse effects to a single cause. Nevertheless, we have discovered that a process long known to discolor and toughen foods may also contribute to age-related impairment of both cells and tissues. That process is the chemical attachment of the sugar glucose to proteins (and, we have found, to nucleic acids) without the aid of enzymes. When enzymes attach glucose to proteins, they do so at a specific site on a specific molecule for a specific purpose. In contrast, the nonenzymatic process adds glucose haphazardly to any of several sites along any available peptide chain.

On the basis of recent in vitro and in vivo studies in our laboratory at Rockefeller University, we propose that this nonenzymatic "glycosylation" of certain proteins in the body triggers a series of chemical reactions that culminate in the formation, and eventual accumulation, of irreversible cross-links between adjacent protein molecules. If this hypothesis is correct, it would help to explain why various proteins, particularly ones that give structure to tissues and organs, become increasingly cross-linked as people age. Although no one has yet satisfactorily described the origin of all such bridges, many investigators agree that extensive cross-linking of proteins probably contributes to the stiffening and loss of elasticity characteristic of aging tissues. We also propose that the nonenzymatic addition of glucose to nucleic acids may gradually damage DNA.

The steps by which glucose alters proteins have been understood by food chemists for decades, although few biologists recognized until recently that the same steps could take place in the body. The nonenzymatic reactions between glucose and proteins, collectively known as the Maillard or browning reaction, may seem complicated, but they are fairly straightforward compared with many biochemical reactions.

They begin when an aldehyde group (CHO) of

glucose and an amino group (NH$_2$) of a protein are attracted to each other. The molecules combine, forming what is called a Schiff base (see Figure 12.1). This combination is unstable and quickly rearranges itself into a stabler, but still reversible, substance known as an Amadori product.

If a protein persists in the body for months or years, some of its Amadori products slowly dehydrate and rearrange themselves yet again—into new glucose-derived structures. These can combine with various kinds of molecules to form irreversible structures we have named advanced glycosylation end products (AGE's) (see Figure 12.2). Most AGE's are yellowish brown and fluorescent and have specific spectographic properties. More important for the body, many are also able to cross-link adjacent proteins.

The precise chemical structure of advanced glycosylation end products and of most AGE-derived cross-links is still not known. Nevertheless, some evidence suggests that AGE's are often created by the binding of an Amadori product to glucose or another sugar. Such end products would form bridges to other proteins by binding to available amino groups. In some instances two Amadori products may instead merge, creating an AGE that is also a cross-link. The one glucose-derived cross-link whose chemical structure is known appears to be just such a combination. It is 2-furanyl-4(5)-(2-furanyl)-1*H*-imidazole, or FFI (see Figure 12.3). First isolated in the laboratory (from a mixture of the amino acid lysine, the protein albumin and glucose), FFI has since been found in the body.

The realization that the browning reaction could occur in—and potentially damage—the body emerged from studies of diabetes, a disease characterized by elevated blood-glucose levels. In the mid-1970's one of us (Cerami) and Ronald J. Koenig examined a report that the blood of diabetic individuals contained higher than normal levels of hemoglobin A$_{1c}$: a variant of the protein hemoglobin,

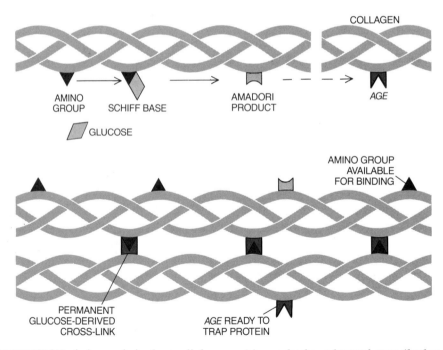

Figure 12.1 FORMATION of glucose-derived cross-links, shown schematically, begins when glucose attaches to an amino group (NH$_2$) of a protein (*top*), such as collagen. The initial product, a Schiff base, soon transforms itself into an Amadori product, which can eventually pass through several incompletely understood steps (*broken arrow*) to become an AGE. In many instances AGE's are like unsprung traps (*red symbol*), poised to snap shut (*bottom*) on free amino groups of any nearby protein and to form cross-links.

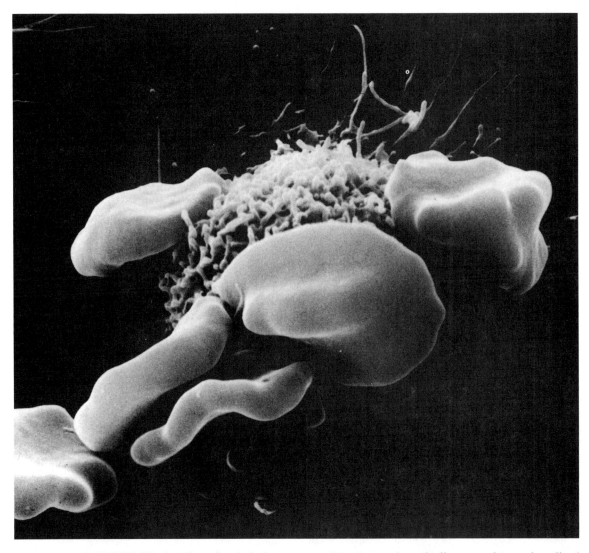

Figure 12.2 MACROPHAGE (*rough-surfaced body at center*), a cell that removes debris from tissues, is about to ingest red blood cells (*smooth disks*) to which advanced glycosylation end products, or AGE's, have been attached. AGE's are molecules derived from the combination of glucose and protein without the aid of enzymes. The authors postulate that AGE's gradually accumulate on long-lived proteins and cells, forming cross-links that impair tissues. Macrophages try to remove AGE-altered proteins but lose efficiency as people age. The cells are enlarged 10,000 diameters in this micrograph.

which is the oxygen-carrying component of red blood cells. Curious about why the levels were elevated, the two investigators attempted to determine the molecule's structure.

Hemoglobin A_{1c} is an Amadori product. Moreover, as is true for the amount of Amadori product formed in foods, the amount of hemoglobin A_{1c} formed is influenced by the level of glucose in the blood: when the glucose level is high, the amount of Amadori product is also high. (Workers in our laboratory and elsewhere have since identified more than 20 Amadori proteins in human beings and have consistently found two or three times as much product in people with diabetes as in nondiabetics.)

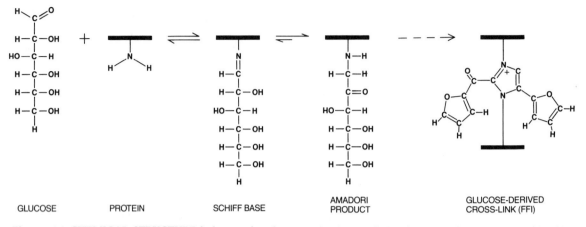

GLUCOSE   PROTEIN   SCHIFF BASE   AMADORI PRODUCT   GLUCOSE-DERIVED CROSS-LINK (FFI)

Figure 12.3 CHEMICAL STRUCTURE is known for glucose-protein Schiff bases and Amadori products. Workers have yet to learn the structure of most AGE's and AGE-derived cross-links, but one link has been identified: 2-furanyl-4(5)-(2-furanyl)-1*H*-imidazole, or FFI.

The hemoglobin findings reveal that glucose, which bathes tissues and cells throughout the body, is not the inert biological molecule most biologists thought it was. Although the sugar does not react while it is in its usual ringlike formation, the ring opens often enough to enable Amadori products and other substances to form. Glucose remains the least reactive sugar in the body, but it has the greatest potential effect on proteins because it is by far the most abundant variety.

The fact that glucose is reactive suggested to Cerami that excess blood glucose in people with uncontrolled diabetes might be more than a marker of the disease. If the sugar could bind nonenzymatically to proteins in the body, he reasoned, excessive amounts could potentially contribute to diabetic complications: the host of disorders, ranging from impaired sensation to kidney failure, that often disable people with diabetes and shorten their life. In particular, it seemed possible that high levels of glucose could lead to an extensive buildup of advanced glycosylation end products on long-lived proteins. The accumulation of AGE's in turn might undesirably modify tissues throughout the body.

Such musings soon led to a suspicion that glucose could also play a role in the tissue changes associated with normal aging. The effect of diabetes on many organs and tissues is often described as accelerated aging because several of the complications that strike people with diabetes—including senile cataracts, joint stiffness and atherosclerosis—are identical with disorders that develop in the elderly; they merely develop earlier. If excess glucose does in fact hasten the onset of these ills in people with diabetes, normal amounts could conceivably play a role in the slower onset seen in nondiabetics as they age.

Our laboratory's studies of senescence (which complement our ongoing studies of diabetes) began with an attempt to determine whether advanced glycosylation end products do in fact accumulate on, and form cross-links between, long-lived proteins in the body. Major constituents of the lens of the eye—the crystallin proteins—became the first objects of study because once these proteins are produced they are believed to persist for life; they therefore fit the profile of proteins that could amass advanced glycosylation end products. Also it seemed likely that a buildup of such AGE's and of AGE-derived cross-links could help to explain why lenses turn brown and cloudy (that is, develop senile cataracts) as people age. In support of this idea, workers elsewhere had previously found two types of cross-link in aggregates of crystallin proteins from human senile cataracts. One bridge was pigmented, suggesting that it could be an AGE. The other type was a disulfide bond formed between sulfhydryl (SH) groups of the amino acid cysteine.

In test-tube experiments Cerami, Victor J. Stevens and Vincent M. Monnier showed that glucose could produce a cataractlike state in a solution of the proteins. Whereas glucose-free solutions containing crystallins from bovine lenses remained clear, solutions with glucose caused the proteins to form clusters, suggesting that the molecules had become cross-linked. The clusters diffracted light, making the solution opaque. Analysis of the links between the molecules confirmed that both the disulfide and the pigmented types were present. The group has also discovered that the pigmented cross-links in human senile cataracts have the brownish color and fluorescence characteristic of advanced glycosylation end products. In fact, some cross-links can be chemically identified as the advanced glycosylation end product FFI.

Combined with other evidence, the above data suggest that nonenzymatic glycosylation of lens crystallins may contribute to cataract formation by a two-step mechanism. Glucose probably alters the conformation of proteins in ways that render previously unexposed sulfhydryl groups susceptible to combination with nearby sulfhydryl groups. Hence disulfide bonds develop, initiating protein aggregation. Later Amadori products on the proteins become rearranged, enabling FFI and other pigmented cross-links to form, discolor the lens and make it cloudy.

Convinced that at least one class of proteins undergoes the browning reaction and forms undesirable cross-links, we and our colleagues turned to the body's most abundant protein: collagen. This long-lived extracellular protein glues together the cells of many organs and helps to provide a scaffolding that shapes and supports blood-vessel walls. It also is a major constituent of tendon, skin, cartilage and other connective tissues. In the past 25 years various investigators have shown that collagen builds up in many tissues, becoming increasingly cross-linked and stiff as people age (see Figure 12.4).

Studies of the dura mater, the collagen sac separating the brain from the skull, provided early evidence that advanced glycosylation end products could collect on collagen. Monnier, Cerami and the late Robert R. Kohn of Case Western Reserve University found that the dura mater from elderly individuals and from diabetics displays yellowish brown pigments whose fluorescent and spectrographic properties are similar to those of advanced glycosylation end products formed in the test tube. As would be expected, protein from people with diabetes had accumulated more pigments than the protein of nondiabetics. In nondiabetics the amount

of pigment attached to the protein increased linearly with age.

Evidence suggesting that glucose induces collagen not only to form AGE's but also to become cross-linked comes from several studies. On the basis of work by other investigators, it has long been known that fibers from the tail tendons of older rats take longer to break when they are stretched than fibers from younger animals, indicating that the older fibers are more cross-linked and less flexible. Monnier, Cerami and Kohn therefore attempted to mimic the effects of aging by incubating tendon fibers of young rats with various sugars. The fibers gradually accumulated advanced glycosylation end products and showed a concomitant increase in breaking time.

More recently we have evaluated the cross-linking of both purified and aortic collagen. In the first instance the protein was incubated with glucose in a test tube; in the second instance it was essentially incubated in the body of diabetic animals that had high blood-glucose levels. In both conditions our chemical tests unequivocally showed that the glucose led to extensive cross-linking.

Although we suspect that the formation of glucose-derived cross-links between long-lived proteins helps to account for many symptoms of aging and for many complications of diabetes, such bridges are not the only ones that can potentially damage the body. We have shown that AGE's on collagen in artery walls and in the basement membrane of capillaries can actually trap a variety of normally short-lived plasma proteins. Even when collagen is incubated with glucose and then washed so that no free glucose is present, the long-lived protein can still covalently bind such molecules as albumin, immunoglobulins and low-density lipoproteins.

This binding may help to explain why both people with diabetes and the aged are prone to atherosclerosis: a buildup of plaque in arterial walls. The plaque includes smooth-muscle cells, collagen (which is produced by the smooth-muscle cells) and lipoproteins (the cholesterol-rich proteins that are the primary source of fat and cholesterol in atherosclerotic lesions).

No one yet understands the exact processes leading to atherosclerosis. It is conceivable that glucose contributes to plaque formation by causing advanced glycosylation end products to develop progressively on collagen in the vessel walls. Once those substances form, collagen may trap low-den-

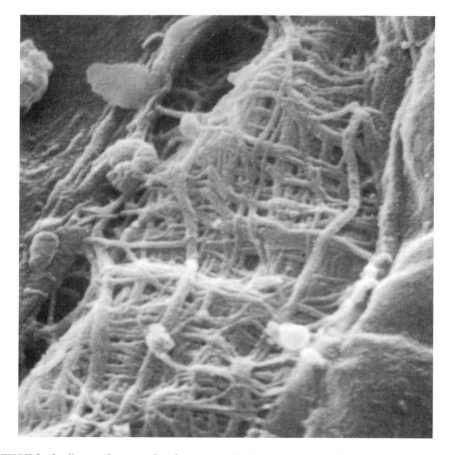

Figure 12.4 FIBRILS of collagen, the most abundant protein in the animal world, are enlarged 26,000 diameters in this micrograph of chick-embryo collagen. As animals and people age, cross-linking of the protein molecules in such fibrils causes tissues throughout the body to stiffen. The exact nature of all cross-links is not known, but evidence suggests that many of them may be AGE-derived.

sity lipoproteins from the blood—which in turn can become attachment sites for other lipoproteins.

In theory, glucose-altered collagen could also trap von Willebrand factor, a protein that is believed to promote the aggregation of platelets (sticky bodies involved in blood clotting). The platelets may release a factor that stimulates the proliferation of smooth-muscle cells, which produce extra collagen. Other glucose-related events may further promote plaque formation (see Figure 12.5). More studies are needed to determine the extent to which any of the postulated events take place and how they might

Figure 12.5 GLYCOSYLATION END PRODUCTS are suspected of contributing to atherosclerosis, and thus to coronary disease, by several pathways. When the inner lining of a healthy blood vessel (*top*) is damaged, plasma proteins leak into the arterial wall (*bottom*). AGE's on collagen in the wall then trap low-density lipoproteins (LDL), which accumulate to form cholesterol deposits in atherosclerotic plaques. Macrophages attempt to remove the captured lipoproteins and in the process secrete a factor that stimulates smooth-muscle cells to proliferate and make new collagen (*blue*). Finally, AGE's on collagen may trap von Willebrand factor, a protein that causes platelets to adhere to the vessel wall. Like macrophages, activated platelets secrete a cell-proliferation factor.

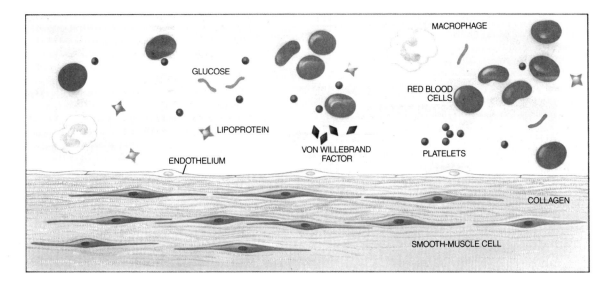

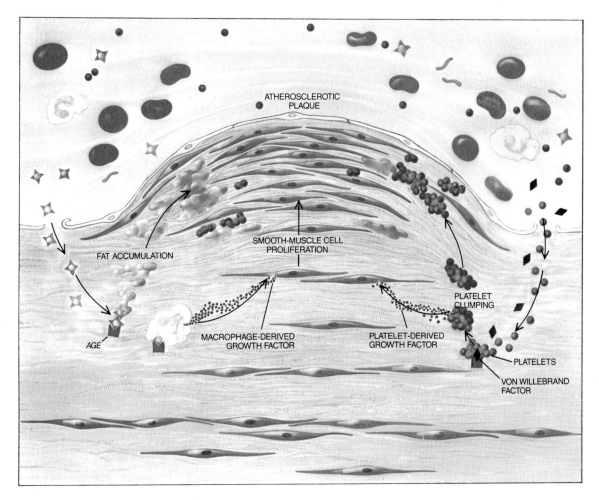

interact with various other processes that contribute to atherosclerosis.

Protein trapping and cross-linking may also help to explain the thickening seen in the basement membrane of capillaries as people grow older (and the more rapid thickening in people with diabetes). In people with diabetes, thickening of a specialized basement membrane in the kidney, the mesangial matrix, promotes renal failure. In nondiabetics the consequences of renal basement-membrane thickening are less clear, although we suspect the process may help to decrease the aged kidney's ability to clear wastes from the blood. Elsewhere in the body thickened capillaries become particularly narrow or occluded in the course of time in the lower extremities, where gravity increases the rate of protein trapping by vessel walls. Such narrowing can contribute to the impaired circulation and loss of sensation often found in the feet and legs of both diabetics and older nondiabetics. In order to function properly, the sensory nerves need an adequate supply of blood.

Because aging takes place at the level of the cell as well as of tissue, our laboratory has recently begun to examine the effects of glucose on the material that controls cell activity: the genes. At least in resting cells, the nucleic acid DNA, which contains amino groups, is long-lived. It therefore can potentially accumulate advanced glycosylation end products. These AGE's might then contribute to known age-related increases in chromosomal alterations or to declines in the repair, replication and transcription of DNA. Such genetic changes are believed to impair the body's ability to replace proteins critical to normal cell function and survival. Nonenzymatic glycosylation might also cause mutations that affect the activity of the immune system or lead to some types of cancer.

Richard Bucala, Peter Model and Cerami have found that incubating DNA with glucose does indeed cause fluorescent pigments to form. The pigments do not build up as quickly as they do on proteins, because the amino groups of nucleic acids are significantly less reactive than the amino groups of proteins.

No one has yet investigated the effects of AGE's on the nucleic acids of mammalian cells, but the group's studies of bacteria suggest that nonenzymatic glycosylation may well interfere with the normal functioning of human genes. When a bacteriophage (a bacterial virus) with a DNA genome was incubated with glucose and then inserted into the bacterium Escherichia coli, the phage's ability to infect E. coli cells was shown to be reduced. The degree of reduction depended on both the incubation time and the concentration of the sugar.

Bucala and his fellow workers also found that adding the amino acid lysine to a mixture of DNA and glucose hastened the loss of viral activity. Presumably the sugar reacted with the amino acid, forming an "AGE-lysine" that quickly bound to the DNA. Because both protein and glucose are present in mammalian cells, it seems likely that a similar reaction might account for the finding that protein covalently attaches to the DNA of aged cells. The effects of such protein binding to genetic material are not known.

Just how the attachment of glucose or a glycosylated protein to DNA interfered with the bacteriophage's normal activity is also not clear. In another study, though, sugar was shown to cause a mutation in DNA. The workers isolated plasmids (extrachromosomal pieces of bacterial DNA) carrying genes that make E. coli resistant to the antibiotics ampicillin and tetracycline (see Figure 12.6). Then they incubated the plasmid with glucose-6-phosphate, a sugar that reacts more quickly than glucose, returned the DNA to bacterial cells and exposed the cells to an antibiotic. Most of the cells exposed to tetracycline died, whereas most exposed to ampicillin lived. Clearly some of the incubated plasmids kept the ampicillin-resistance gene but had lost the activity of the tetracycline-resistance gene.

Further study showed that most of the tetracycline-resistance genes had been altered by deletions or insertions of DNA. We suspect that those genes had collected advanced glycosylation end products and that the resulting mutations arose when the bacteria attempted to repair the DNA altered by AGE's. This conclusion is supported by the finding that bacterial cells lacking a DNA-repair enzyme did not have mutations in the DNA.

In order to better determine the effects of advanced glycosylation end products on the DNA of human cells, we are developing new methods for measuring both AGE's and glycosylated proteins on DNA. In addition we need to learn more about the cell's mechanisms for repairing glycosylated nucleic acids.

The ultimate goal of our research into both aging and diabetes is to find ways of preventing or delaying their debilitating effects. If our glycosyla-

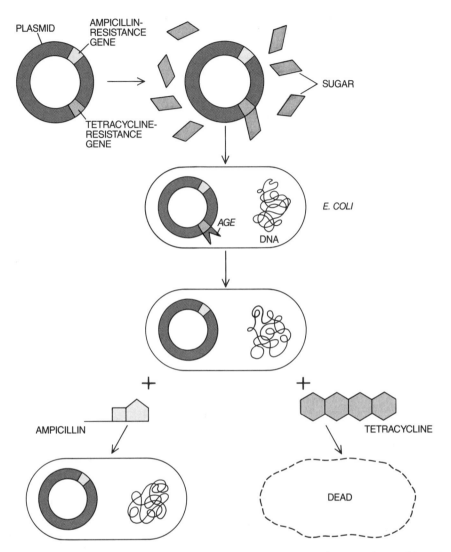

Figure 12.6 PLASMID, an extrachromosomal circle of bacterial DNA, underwent mutation as a result of being incubated with sugar, suggesting that glucose may contribute to some of the genetic damage seen in humans as they age. After incubation, plasmids carrying genes that render the bacterium *Escherichia coli* resistant to the antibiotics ampicillin and tetracycline were inserted into *E. coli*. In the presence of ampicillin the bacterial cells reproduced normally, but in the presence of tetracycline most cells died. Apparently bacterial enzymes attempted to repair tetracycline-resistance genes that accumulated AGE's.

tion hypothesis is correct, such effects might be mitigated either by preventing the formation of glucose-derived cross-links or by increasing the activity of biological processes that remove AGE's.

On the first front we, along with Peter C. Ulrich in our laboratory, have developed a promising drug called aminoguanidine (see Figure 12.7). This small molecule, in the class of compounds called hydrazines, reacts with Amadori products. It apparently binds to carbonyl groups and in so doing prevents the Amadori products from becoming advanced glycosylation end products.

In test-tube studies of the drug we incubated albumin either with glucose alone or with glucose and

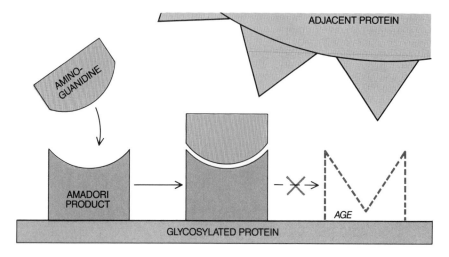

Figure 12.7 AMINOGUANIDINE is an experimental drug that interferes with the ability of Amadori products to undergo changes that could normally result in the forma-tion of cross-links. The drug's safety and efficacy in humans are under study.

aminoguanidine. Advanced glycosylation end products formed in the first mixture within half a week and increased progressively with time. In contrast, the aminoguanidine mixture produced an equal amount of Amadori product but resulted in marked inhibition of AGE formation. Similarly, when we incubated collagen with glucose, the protein became extremely cross-linked, whereas the addition of aminoguanidine blocked nearly all glucose-derived intermolecular bridges.

Parallel findings come from studies of diabetic rats. Animals treated with aminoguanidine amassed fewer advanced glycosylation end products in the aorta, and fewer cross-links, than untreated rats. In a separate group of diabetic rats we have shown that aminoguanidine prevents both the trapping of immunoglobulin in the basement membrane of renal capillaries and the trapping of plasma lipoproteins in the arterial wall.

We are now planning trials of aminoguanidine in human subjects. If the drug is shown to be safe, we hope to conduct long-term trials of its ability to prevent diabetic complications. Because diabetes is in some ways a model of aging, success in such trials might eventually help to justify studying the ability of aminoguanidine (or similar compounds) to prevent disorders related to age in nondiabetics.

We are also studying the other approach to treatment: increasing the activity of the body's AGE-removal system. Even if the production of advanced glycosylation end products could not be prevented, an effective AGE-removal system might help to counteract any dangerous buildup on proteins. Macrophages, the scavenger cells that remove debris from tissues, apparently constitute one such removal system.

This property of the scavenger cells became clear about three years ago when we examined peripheral-nerve myelin: the complex mixture of long-lived proteins that forms an insulting sheath around nerve fibers. We incubated isolated myelin with glucose for eight weeks to mimic the effects of long-term exposure to glucose in the body. Then we introduced macrophages into the mixture. The cells ingested more myelin than they did when the substance had not been exposed to sugar. They also took up more myelin from diabetic animals than from nondiabetic ones, presumably because the diabetic animals had a greater amount of advanced glycosylation end products. (See Figure 12.2.)

More recent evidence indicates that the signal for protein uptake by macrophages is specifically the advanced glycosylation end product. We have found, for example, that a mouse macrophage has an estimated 150,000 receptors for the AGE's that form on albumin. Macrophages attempt to ingest any protein attached to the advanced glycosylation end product FFI, but the cells' receptors do not appear to react with any non-AGE substances that accumulate on proteins, including Amadori products.

The affinity of macrophages for FFI, and for advanced glycosylation end products in general, became dramatically apparent when we attached FFI and other AGE's to membrane proteins of normal red blood cells. Mouse macrophages took up the altered cells much more avidly than they take up normal cells. (In addition to supporting the contention that macrophages are an AGE-removal system, this discovery suggests that advanced glycosylation end products have at least one constructive role in the body: they may indicate that a cell is aged and should be removed.)

Why do AGE's build up on proteins if the body has a system for removing them? We do not have an answer, but a few explanations seem likely. For one thing, the end products may generally form in locations that are not readily accessible to macrophages. Moreover, the highly cross-linked proteins that eventually accumulate appear to be increasingly difficult to remove. Also, as people age, their macrophages may become less efficient as a disposal

mechanism. In support of this last notion we have very recently discovered that the number of AGE receptors on mouse macrophages declines as mice grow older.

We are currently seeking drugs that increase the removal rate of unwanted advanced glycosylation end products, but a successful treatment will have to dissolve the end products without excessively damaging irreplaceable proteins. In the case of myelin, for instance, excess AGE-stimulated uptake of old or damaged protein could erode the myelin sheath, which is essential to nerve functioning.

Additional evidence must be gathered before we can say with certainty that nonenzymatic glycosylation of proteins contributes to the cell and tissue changes characteristic of aging. The data collected so far do indicate that our hypothesis is a promising one. More important, the findings raise the exciting possibility that treatments can one day be developed to prevent some of the changes that too often make "aging" synonymous with "illness."

How LDL Receptors Influence Cholesterol and Atherosclerosis

LDL receptors bind particles carrying cholesterol and remove them from the circulation. Many Americans have too few LDL receptors, and so they are at high risk for atherosclerosis and heart attacks.

. . .

Michael S. Brown and Joseph L. Goldstein
November, 1984

Half of all deaths in the U.S. are caused by atherosclerosis, the disease in which cholesterol, accumulating in the wall of arteries, forms bulky plaques that inhibit the flow of blood until a clot eventually forms, obstructing an artery and causing a heart attack or a stroke (see Figure 13.1). The cholesterol of atherosclerotic plaques is derived from particles called low-density lipoprotein (LDL) that circulate in the bloodstream. The more LDL there is in the blood, the more rapidly atherosclerosis develops (see Figure 13.2).

Epidemiologic data reveal the surprising fact that more than half of the people in Western industrialized societies, including the U.S., have a level of circulating LDL that puts them at high risk for developing atherosclerosis. Because such concentrations are so prevalent, they are considered "normal," but clearly they are not truly normal. They predispose to accelerated atherosclerosis and heart attacks or strokes.

What determines the blood level of LDL, and why is the level dangerously high in so many Americans? Some answers are emerging from studies of specialized proteins, called LDL receptors, that project from the surface of animal cells (see Figure 13.3). The receptors bind LDL particles and extract them from the fluid that bathes the cells. The LDL is taken into the cells and broken down, yielding its cholesterol to serve each cell's needs. In supplying cells with cholesterol the receptors perform a second physiological function, which is critical to the development of atherosclerosis: they remove LDL from the bloodstream.

The number of receptors displayed on the surface of cells varies with the cells' demand for cholesterol. When the need is low, excess cholesterol accumulates; cells make fewer receptors and take up LDL at a reduced rate. This protects cells against excess cholesterol, but at a high price: the reduction in the number of receptors decreases the rate at which LDL is removed from the circulation, the blood level of LDL rises and atherosclerosis is accelerated.

We have proposed that the high level of LDL in many Americans is attributed to a combination of factors that diminish the production of LDL receptors. Recognition of the central role of the receptors has led to a treatment for a severe genetic form of atherosclerosis, and it has also shed some light on

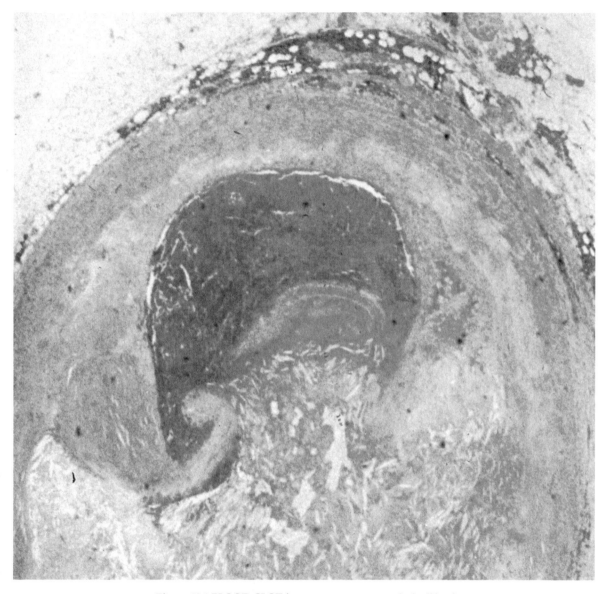

Figure 13.1 BLOOD CLOT in a coronary artery took the life of a 76-year-old man with advanced atherosclerosis. Over the decades cholesterol carried by low-density lipoprotein (LDL) particles had infiltrated the wall of the artery, forming a bulky deposit (*pale pink*) that narrowed the channel. The clot formed suddenly, obstructing blood flow and causing the death of the heart muscle that had been supplied with oxygen and nutrients by this artery. The formation of the clot (*dark red*) appears to have been triggered by a rupture of the lining of the channel that exposed the flowing blood to crystals of cholesterol (*white needle-shaped objects*). The sectioned artery is enlarged 37 diameters.

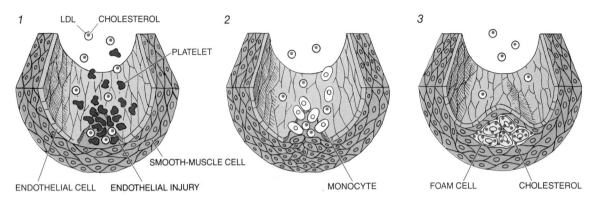

Figure 13.2 ATHEROSCLEROTIC PLAQUE develops slowly. Damage to the thin layer of endothelial cells that lines an artery initiates plaque formation. According to a model proposed by Russell Ross and John A. Glomset, the damaged endothelium becomes leaky and is penetrated by low-density lipoprotein (LDL) particles and blood platelets (*1*). In response to the release of such hormones as platelet-derived growth factor, smooth-muscle cells in the layer below the endothelium multiply and migrate into the damaged area (*2*); at the same time monocytes invade the area and are activated to become macrophages. The smooth-muscle cells and macrophages ingest and degrade LDL and become foam cells. If the blood LDL level is too elevated, cholesterol derived from the LDL accumulates in and among the foam cells. The accumulated cholesterol, cells and debris constitute an atheroma (*3*), which in time can narrow the channel of the artery and led to thrombosis.

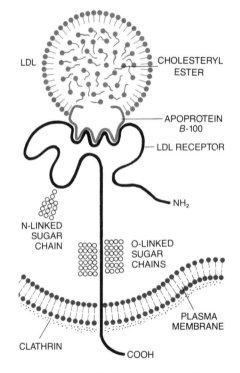

Figure 13.3 LDL RECEPTOR, a glycoprotein embedded in the plasma membrane of most body cells, was purified from the adrenal gland by Wolfgang J. Schneider. David W. Russell and Tokuo Yamamoto cloned complementary DNA derived from its messenger RNA. The DNA's nucleotide sequence was determined and from it the 839-amino-acid sequence of the receptor's protein backbone was deduced. Sites of attachment of sugar chains to nitrogen (N) and oxygen (O) atoms were identified, as was a stretch likely to traverse the membrane. The actual shape of the receptor is not yet known; the drawing is a highly schematic representation.

the continuing controversy over the role of diet in atherosclerosis in the general population.

The story begins with the discovery of LDL receptors in 1973 in our laboratory at the University of Texas Health Science Center at Dallas. We were studying tissue cultures of the human skin cells called fibroblasts. Like all animal cells, cultured fibroblasts need cholesterol as a major building block of their surface membrane (the plasma membrane); they had been shown to get the cholesterol by extracting it from lipoproteins in the serum of the culture medium. There is a mixture of various lipo-

proteins in human serum, but we found that the fibroblasts derive most of their cholesterol from a particular lipoprotein: LDL. We were able to attribute this to the presence on the cells of highly specific receptor molecules that bind LDL and related lipoproteins.

LDL is a large spherical particle whose oily core is composed of some 1,500 molecules of the fatty alcohol cholesterol, each attached by an ester linkage to a long-chain fatty acid (see Figure 13.4). This core of cholesteryl esters is enclosed in a layer of phospholipid and unesterified cholesterol molecules. The phospholipids are arrayed so that their hydrophilic

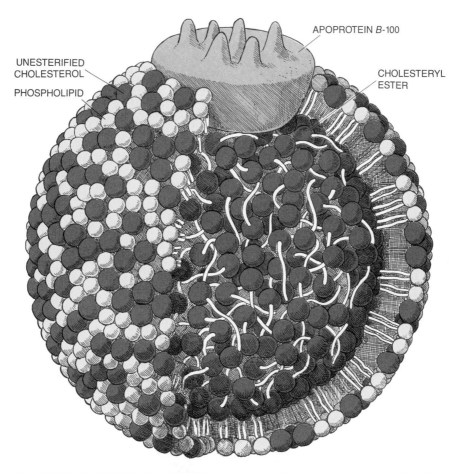

UNESTERIFIED CHOLESTEROL

PHOSPHOLIPID

APOPROTEIN *B*-100

CHOLESTERYL ESTER

Figure 13.4 LDL, MAJOR CHOLESTEROL CARRIER in the bloodstream, is a spherical particle with a mass of three million daltons and a diameter of 22 nanometers (millionths of a millimeter). Its core consists of some 1,500 cholesteryl esters, each a cholesterol molecule attached by an ester linkage to a long fatty acid chain. The oily core is shielded from the aqueous plasma by a detergent coat composed of 800 molecules of phospholipid, 500 molecules of unesterified cholesterol and one large protein molecule, apoprotein *B*-100. When blood cholesterol is elevated, increasing the risk of atherosclerosis, the reason is almost always an increase in circulating LDL.

heads are on the outside, allowing the LDL to be dissolved in the blood or intercellular fluid. Embedded in this hydrophilic coat is one large protein molecule designated apoprotein B-100.

It is apoprotein B-100 that is recognized and bound by the LDL receptor, a glycoprotein (a protein to which sugar chains are attached). The receptor spans the thickness of the cell's plasma membrane and carries a binding site that protrudes from the cell surface. Binding takes place when LDL is present at a concentration of less than 10^{-9} molar, which is to say that the receptor can pick out a single LDL particle from more than a billion molecules of water. The receptor binds only lipoproteins carrying apoprotein B-100 or a related protein designated apoprotein E.

How is LDL taken into the cell? Our collaborator Richard G. W. Anderson discovered in 1976 that the receptors are clustered in specialized regions where the cell membrane is indented to form craters known as coated pits (because the inner surface of the membrane under them is coated with the protein clathrin). Within minutes of their formation the pits pouch inward into the cell and pinch off from the surface to form membrane-bounded sacs called coated vesicles; any LDL bound to a receptor is carried into the cell. (Receptor-mediated endocytosis, the term we and Anderson applied to this process of uptake through coated pits and vesicles, is now recognized as being a general mechanism whereby cells take up many large molecules, each having its own highly specific receptor.)

Eventually the LDL is separated from the receptor (which is recycled to the cell surface) and is delivered to a lysosome, a sac filled with digestive enzymes. Some of the enzymes break down the LDL's coat, exposing the cholesteryl ester core. Another enzyme clips off the fatty acid tails of the cholesteryl esters, liberating unesterified cholesterol, which leaves the lysosome. As we have indicated, all cells incorporate the cholesterol into newly synthesized surface membranes. In certain specialized cells the cholesterol extracted from LDL has other roles. In the adrenal gland and in the ovary it is converted into respectively the steroid hormones cortisol and estradiol: in the liver it is transformed to make bile acids, which have a digestive function in the intestine.

The amount of cholesterol liberated from LDL controls the cell's cholesterol metabolism (see Figure 13.5). An accumulation of cholesterol modulates three processes. First, it reduces the cell's ability to make its own cholesterol by turning off the synthesis of an enzyme, HMG CoA reductase, that catalyzes a step in cholesterol's biosynthetic pathway. Suppression of the enzyme leaves the cell dependent on external cholesterol derived from the receptor-mediated uptake of LDL. Second, the incoming LDL-derived cholesterol promotes the storage of cholesterol in the cell by activating an enzyme called ACAT. The enzyme reattaches a fatty acid to excess cholesterol molecules, making cholesteryl esters that are deposited in storage droplets.

Third, and most significant, the accumulation of cholesterol within the cell drives a feedback mechanism that makes the cell stop synthesizing new LDL receptors. Cells thereby adjust their complement of receptors so that enough cholesterol is brought in to meet their varying demands but not enough to overload them. For example, fibroblasts that are actively dividing, so that new membrane material is needed, maintain a maximum complement of LDL receptors (some 40,000 per cell). In cells that are not growing the incoming cholesterol begins to accumulate, the feedback system reduces receptor manufacture and the complement of receptors is reduced as much as tenfold.

Our observations in tissue cultures were confirmed when the receptor system was shown to have an important role in the body. Soon after we found the LDL receptor on cultured fibroblasts it was shown to be present on circulating human blood cells and on cell membranes from many different tissues of mice, rats, dogs, pigs, cows and human beings. The relative number of receptors and their functioning can be assessed in living animals and in human volunteers by injecting into the bloodstream LDL labeled with a radioactive isotope and measuring its rate of removal from the circulation. The rate has been shown to depend on the total number of LDL receptors displayed on all cells in the body. This can be demonstrated by modifying the apoprotein B-100 before the LDL is injected, so that it can no longer bind to receptors. James Shepherd and Christopher J. Packard of the University of Glasgow showed that the modified LDL circulates much longer than normal LDL.

Where is the LDL taken up? Daniel Steinberg of the University of California School of Medicine at San Diego and John M. Dietschy of the Health Science Center at Dallas have shown that in rats, rabbits, guinea pigs and squirrel monkeys about 75 percent of the receptor-mediated removal of LDL takes place in the liver. We have measured the number of receptors directly, in cell membranes isolated

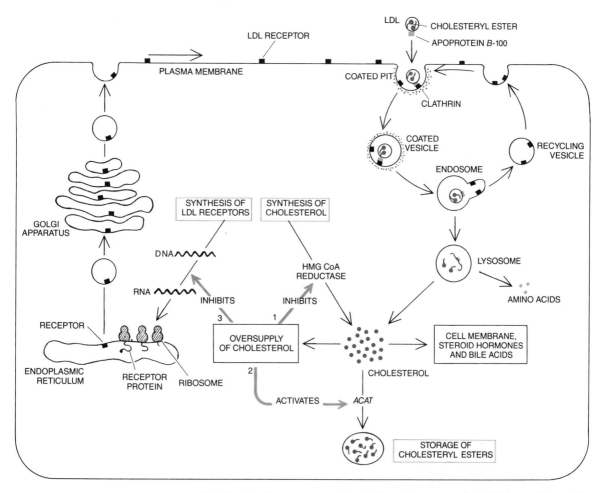

Figure 13.5 CIRCULATING LDL (*top right*) is taken into a cell by receptor-mediated endocytosis. LDL is bound by a receptor in a coated pit, which invaginates and pinches off to form a coated vesicle. Fusion of several vesicles gives rise to an endosome, in whose acidic environment the LDL dissociates from the receptor, which is recycled to the cell surface. The LDL is delivered to a lysosome, where en- **zymes break down the apoprotein *B*-100 into amino acids and cleave the ester bond to yield unesterified cholesterol. An oversupply of cholesterol inhibits the enzyme HMG CoA reductase (*1*); activates the enzyme *ACAT*, (*2*), and inhibits the manufacture of new LDL receptors by suppressing transcription of the receptor gene into messenger RNA (*3*).**

from different tissues. Most tissues are found to have some receptors, but those of the liver, adrenal gland and ovary — the organs with particularly large requirements for cholesterol — have the highest concentration of receptors.

What is the origin of circulating LDL? The mechanism of its production is more complex, and as yet less well understood, than the mechanism of its uptake and degradation. LDL is one component of the system that transports two fatty substances, cholesterol and various triglycerides, through the bloodstream. The fat-transport system can be divided into two pathways: an exogenous one for cholesterol and triglyceride absorbed from the intestine and an endogenous one for cholesterol and triglyceride entering the bloodstream from the liver and other nonintestinal tissues (see Figure 13.6).

The exogenous pathway has been mapped by Richard J. Havel of the University of California School of Medicine at San Francisco and by others.

Figure 13.6 EXOGENOUS AND ENDOGENOUS fat-transport pathways. Cholesterol is absorbed through the wall of the intestine and is packaged with triglyceride in chylomicrons. In the capillaries of fat and muscle tissue the triglyceride's ester bond is cleaved by lipoprotein (*LP*) lipase and the fatty acids are removed. When the cholesterol-rich remnants reach the liver, they bind to receptors and are taken into liver cells. Their cholesterol either is secreted into the intestine or is packaged with triglyceride in VLDL particles and secreted into the circulation, inaugurating the endogenous pathway. The triglyceride is removed in fat or muscle, leaving cholesterol-rich IDL. Some IDL binds to liver LDL receptors and is rapidly taken up by liver cells; the remainder stays in the circulation and is converted into LDL. Most of the LDL binds to LDL receptors on liver or other cells and is removed from the circulation. Cholesterol leaching from cells binds to HDL and is esterified by the enzyme *LCAT*. The esters are transferred to IDL and then LDL and are eventually taken up again by cells.

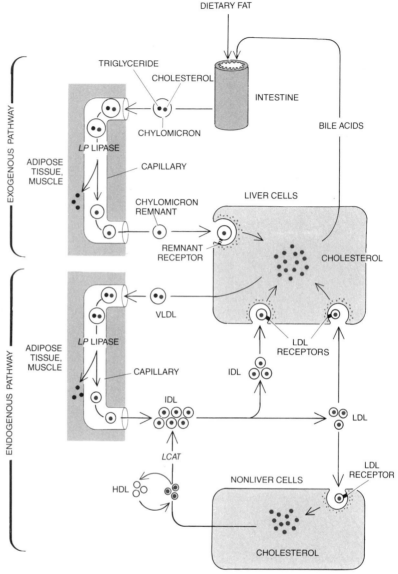

It begins in the intestine, where dietary fats are packaged into lipoprotein particles called chylomicrons, which enter the bloodstream and deliver their triglyceride to adipose tissue (for storage) and to muscle (for oxidation to supply energy). The remnant of the chylomicron, containing cholesteryl esters, is removed from the circulation by a specific receptor found only on liver cells. This chylo-

micron-remnant receptor does not bind LDL or take part in its removal from the circulation.

LDL is a component of the endogenous pathway, which begins when the liver secretes into the bloodstream a large very-low-density lipoprotein particle (VLDL). Its core consists mostly of triglyceride synthesized in the liver, with a smaller amount of cholesteryl esters: it displays on its surface two predominant proteins, apoproteins B-100 and E, both of which can be bound by LDL receptors. When a VLDL particle reaches the capillaries of adipose tissue or of muscle, its triglyceride is extracted. The result is a new kind of particle, decreased in size and enriched in cholesteryl esters but retaining its two apoproteins: it is called intermediate-density lipoprotein, or IDL.

In human beings about half of the IDL particles are removed from the circulation quickly—within from two to six hours of their formation—because they bind very tightly to liver cells, which extract their cholesterol to make new VLDL and bile acids. Robert W. Mahley and Thomas L. Innerarity of the University of California School of Medicine at San Francisco have shown that the tight binding is attributable to apoprotein E, whose affinity for LDL receptors on liver cells is greater than that of apoprotein B-100. IDL particles not taken up by the liver remain in the circulation much longer. In time the apoprotein E is dissociated from them, leaving the particles, now converted into low-density lipoprotein (LDL), with apoprotein B-100 as their sole protein. Because of B-100's lower affinity for LDL receptors, the LDL particles have a much longer life span than IDL particles: they circulate for an average of two and a half days before binding to LDL receptors in the liver and in other tissues.

The central role of the LDL receptor in atherosclerosis was first appreciated when we showed that its absence is responsible for the severe disease called familial hypercholesterolemia (FH). In 1939 Carl Müller of the Oslo Community Hospital in Norway identified the disease as an inborn error of metabolism causing high blood cholesterol levels and heart attacks in young people; he recognized that it is transmitted as a dominant trait determined by a single gene. In the 1960's Avedis K. Khachadurian at the American University in Beirut and Donald S. Frederickson at the U.S. National Heart and Lung Institute showed there are two forms of the disease, a heterozygous form and a more severe homozygous form. Heterozygotes, who inherit one mutant gene, are quite common: about one in 500 people in most ethnic groups. Their plasma LDL level is twice the normal level (even before birth) and they begin to have heart attacks by the time they are 35; among people under 60 who have heart attacks, one in 20 has heterozygous FH.

If two FH heterozygotes marry (one in 250,000 marriages), each child has one chance in four of inheriting two copies of the mutant gene, one from each parent. Such FH homozygotes (about one in a million people) have a circulating LDL level more than six times higher than normal; heart attacks can occur at the age of two and are almost inevitable by the age of 20. It is notable that these children have none of the risk factors for atherosclerosis other than an elevated LDL level. They have normal blood pressure, do not smoke and do not have a high blood glucose level. Homozygous FH is a vivid experiment of nature. It demonstrates unequivocally the causal relation between an elevated circulating LDL level and atherosclerosis.

By what mechanism is the LDL level elevated? What is the particular function of the mutant gene? When we looked at cultured skin fibroblasts and circulating blood cells from FH homozygotes, we saw that the cells have either no functional LDL receptors at all or very few and therefore cannot bind, internalize and degrade LDL efficiently. The defective gene, in other words, encodes the protein of the LDL receptor. Homozygotes, having inherited two defective receptor genes, cannot synthesize any normal receptors. The cells of FH heterozygotes have one normal receptor gene and one mutant gene; they synthesize half the normal number of receptors and can therefore bind, internalize and degrade LDL at half the normal rate.

Although all FH patients studied to date have a mutation in the gene encoding the LDL receptor, the mutations are not always the same. Depending on the particular site that has undergone mutation, the receptor may not be synthesized at all or it may be synthesized but then fail to be transported to the cell surface, fail to bind LDL or fail to cluster in coated pits.

Studies with radioactively labeled LDL show that the particles survive in the bloodstream of FH homozygotes about two and a half times as long as they do in people with a normal LDL-receptor gene (see Figure 13.7). (Eventually the LDL is removed from the circulation by alternate but much less efficient pathways.) The predictable slowdown in the removal and breakdown of LDL is one major reason for the extremely high LDL level characteristic of FH, but it does not account for the entire rise. In addition to degrading LDL more slowly, a person homozygous for FH actually produces about twice as much LDL per day as a normal person. How can a defect in the LDL receptor lead to the overproduction of LDL? The answer to this question came from

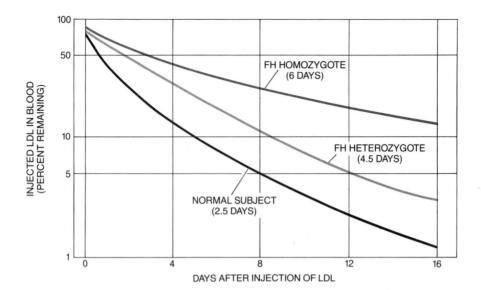

Figure 13.7 NUMBER OF LDL RECEPTORS in the body is assessed by injecting LDL labeled with a radioactive isotope and measuring the amount of radioactivity in blood samples for several weeks; the loss of radioactivity reflects the cellular uptake of LDL and hence the number of LDL receptors. The curves trace the removal of LDL from the circulation in patients with the homozygous and heterozygous forms of familial hypercholesterolemia (FH) and in normal subjects. In each case the mean life span of an LDL particle is shown in parentheses.

markable strain of rabbits with a genetic defect resembling the one in human FH.

The rabbits were discovered in 1978 by Yoshio Watanabe of the Kobe University School of Medicine and are called WHHL rabbits (for "Watanabe heritable hyperlipidemic"). They are homozygous for a mutant LDL-receptor gene and produce less than 5 percent of the normal number of receptors; they have high circulating LDL from the time of birth and develop atherosclerosis leading to heart attacks by the age of two. Studies done by us in collaboration with Toru Kita and David W. Bilheimer and by Steinberg and his colleagues showed that the rabbits, like their human counterparts with homozygous FH, make too much LDL as well as taking too long to break it down.

To learn the reason for LDL overproduction, Kita injected radioactively labeled VLDL, a precursor of LDL, into WHHL rabbits and normal animals and tracked the radioactivity through the fat-transport pathway. He found that triglyceride was removed from the VLDL, generating IDL, at the same rate in both groups. In normal rabbits the vast majority of the IDL particles disappeared rapidly from the circulation as they bound to LDL receptors on liver

cells. In the WHHL rabbits, however, the liver cells lack LDL receptors, and so more IDL particles remained in the circulation and were eventually converted into more than the normal amount of LDL. In other words, a reduction in receptors has two effects in the rabbits—increased production and decreased removal of LDL—that act synergistically to raise the LDL level, which therefore rises disproportionately (see Figure 13.8). Nicholas B. Myant and his colleagues at Hammersmith Hospital in London have shown the same thing is true in FH homozygotes.

K nowledge of the receptor deficiency in FH suggested a way to help the large number of patients with the heterozygous form of the disease. Perhaps we could stimulate the heterozygote's one normal gene to direct the synthesis of twice as many receptors as usual and so provide the patient with a normal complement of functional receptors. The possibility of such treatment was raised by something we had learned from cultured skin fibroblasts, namely that the feedback regulation of receptor synthesis takes place at the level of transcription. An excess of cholesterol reduces transcription of the

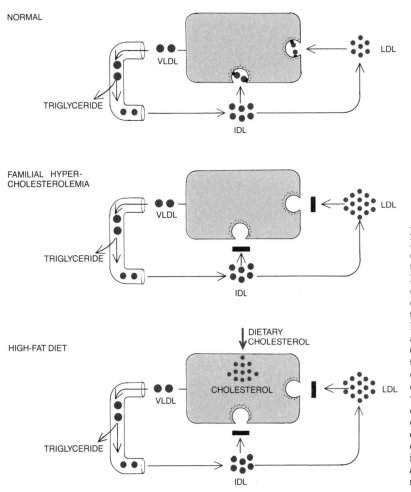

Figure 13.8 LDL-RECEPTOR DE-
FICIENCY, whether genetic or acquired, has two synergistic effects
that combine to raise the blood LDL
level. VLDL secreted by the liver is
converted into IDL in fat and muscle. In normal people about half of
the IDL particles are taken up by
LDL receptors on liver cells; the rest
are converted into LDL (*top*). In FH
(*middle*) a genetic defect diminishes
the number of receptors on liver
cells; an analogous deficiency is
caused by diets that fill liver cells
with cholesterol and so reduce receptor synthesis (*bottom*). In either
case there are the same two consequences. IDL not taken up by liver
cells remains in the circulation and
is converted to yield increased
amounts of LDL; the LDL in turn is
removed more slowly.

LDL-receptor gene into messenger RNA, the nucleic
acid that is subsequently translated by the cell's
protein-synthesizing machinery to make the receptor protein; a cholesterol deficiency stimulates transcription and thus steps up the manufacture of receptors. We found we could get cultured cells from
FH heterozygotes to make a normal number of LDL
receptors (by making more messenger-RNA molecules from their single receptor gene) when we reduced the amount of cholesterol in the culture medium. How might we create an analogous
cholesterol deficiency in the FH patient?

The liver takes up and degrades more cholesterol
than any other organ because of its large size and its
high concentration of LDL receptors. The bile acids
into which most of the cholesterol is converted are

secreted into the upper intestine, where they emul-
sify dietary fats. Having done their work, the bile
acids are not simply excreted, however; they are
largely reabsorbed from the intestine, returned to
the bloodstream, taken up by the liver and again
secreted into the upper intestine. This recycling of
bile acids ordinarily limits the liver's need for cholesterol. We reasoned that if the recycling could be
interrupted, the liver would be called on to convert
more cholesterol into bile acids and this should lead
the liver cells to make more LDL receptors.

A class of drugs that interrupt the recycling of bile
acids was already known. They are the bile-acid-
binding resins, gritty polymers carrying many posi-
tively charged chemical groups. Taken orally, these
resins bind to the negatively charged bile acids in

the intestine; because the resins cannot be absorbed from the intestine, they are excreted, carrying the bound bile acids with them. The first bile-acid-binding resin, cholestyramine, was synthesized more than 20 years ago and was found to lower the blood LDL level by an average of 10 percent. (A recent 10-year prospective study done by the National Heart, Lung, and Blood Institute indicated that such a reduction was enough to cut the incidence of heart attacks in a test group of middle-aged men by 20 percent.) What we had learned about LDL metabolism provided the missing rationale for such results: the interruption of bile-acid recycling increases the number of LDL receptors on liver cells.

The 10 percent drop in LDL level attainable with cholestyramine and other such resins was encouraging, but clearly a more profound reduction is necessary for treating FH heterozygotes. The limited efficacy of the resins stems from the dual response of the liver to a cholesterol deficiency. In addition to making more LDL receptors the liver increases its manufacture of HMG CoA reductase and makes more of its own cholesterol. We reasoned that this increased de novo synthesis of cholesterol partially satisfies the resin-induced demand for more cholesterol and so prevents the liver from maximally increasing the number of LDL receptors.

We thought inhibition of cholesterol synthesis might force the liver to rely more on LDL uptake and thus stimulate greater production of receptors. To block cholesterol synthesis we took advantage of the discovery by Akira Endo, now of the Tokyo University of Agriculture and Technology, of a remarkable natural inhibitor of HMG CoA reductase. In 1976 he isolated from a penicillin mold a substance called compactin. A side chain of the compactin molecule closely mimics the structure of the natural substrate of HMG CoA reductase, and so it binds to the enzyme's active site and inhibits the enzyme's activity. Alfred W. Alberts of the Merck Sharp & Dohme Research Laboratories and his colleagues isolated from a different mold a structural relative of compactin, called mevinolin, that is an even more potent enzyme blocker. Compactin and mevinolin were shown, by Endo and Alberts respectively, to lower the blood LDL level in animals. If our idea was correct, the drugs should be even more effective in conjunction with a bile-acid-binding resin (see Figure 13.9).

In collaboration with Petri T. Kovanen we administered a bile-acid-binding resin to dogs either alone or along with one of the enzyme inhibitors. After two weeks we assessed the number of LDL receptors by measuring the ability of biopsied liver membranes to bind radioactive LDL. We found, as expected, that the resin alone generated a modest rise in the number of receptors. When the enzyme inhibitor was given too, the number of receptors rose much more. At the whole-body level this led to a marked increase in the rate of removal of LDL from the circulation. Together the two drugs caused a remarkable 75 percent decline in the dogs' LDL level.

With Bilheimer and Scott M. Grundy we went on to administer a resin and mevinolin to patients with heterozygous FH (see Figure 13.10). Their LDL level fell by approximately 50 percent, into the normal range. Tests with radioactive LDL showed the drop was caused by an increase in LDL receptors. The single normal gene had been made to work twice as hard as usual, producing enough receptors to allow LDL to be removed from the circulation at a normal rate.

As might be expected, FH homozygotes, lacking even one normal receptor gene, do not respond to this two-drug treatment. Another approach must be found if they are to be helped. Thomas E. Starzl of the University of Pittsburgh School of Medicine has tested a surgical approach, following up on a suggestion that the homozygote's lack of receptors might be partially corrected if the patient could be given a liver from a normal donor. He transplanted the liver of a child killed in an accident into a six-year-old girl suffering from severe homozygous FH. (The patient had already had several heart attacks and her heart was so weakened that a heart transplant was necessary at the same time). More than six months after the operation the patient was maintaining a total blood cholesterol level in the range of 300 milligrams per deciliter, compared with a preoperation level of about 1,200. Obviously liver transplantation is not an ideal treatment, but the results to date make it clear that receptors on the cells of the transplanted liver are functioning to remove LDL from the circulation.

What about the vast number of people in Western industrial societies who suffer heart attacks or strokes without having any genetic defect in the LDL receptor? Is what we have learned about FH relevant to the high incidence of atherosclerosis in the general population? We believe it is (see Figure 13.11). The LDL-receptor hypothesis states that

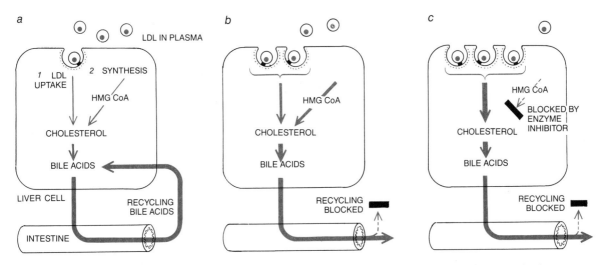

Figure 13.9 LIVER GETS CHOLESTEROL for conversion into bile acids from IDL and LDL taken up from the circulation (*1*) or by synthesizing it de novo (*2*). A key step in the long synthetic pathway is reduction of HMG CoA to mevalonic acid, a reaction catalyzed by the enzyme HMG CoA reductase. The enzyme is inhibited by the drugs compactin or mevinolin, whose side chain is so similar to that of HMG CoA (*colored frames*) that it blocks the enzyme's active site. Enzyme inhibition leaves liver dependent on uptake of IDL and LDL.

much of the atherosclerosis in the general population is caused by a dangerously high blood level of LDL resulting from failure to produce enough LDL receptors. The inadequate number of receptors can be attributed to subtle genetic and environmental factors that limit receptor manufacture even in people without FH. One environmental factor is a high dietary intake of cholesterol and of saturated fats derived from animal tissues.

Epidemiologic surveys done in many countries over the past 30 years have uniformly shown that atherosclerosis becomes severer as the mean LDL

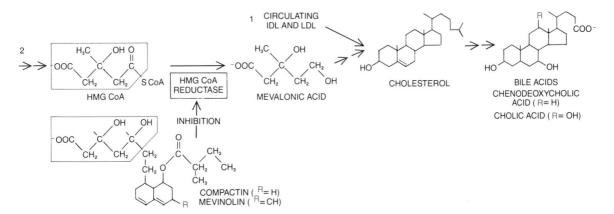

Figure 13.10 HETEROZYGOUS FH can be treated with a combination of drugs that stimulates manufacture of LDL receptors. Ordinarily the liver's demand for cholesterol is modified by the recirculation of bile acids (*a*). If the recirculation is prevented by a bile-acid-binding resin (*b*), more cholesterol is needed. Liver cells respond by increasing the number of LDL receptors, but also by increasing the rate of cholesterol synthesis. If a second drug is given to block enhanced synthesis (*c*), still more receptors are made and the blood LDL level is lowered.

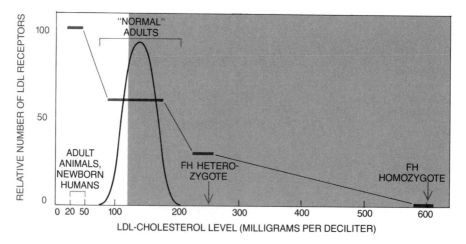

Figure 13.11 RANGE OF LDL LEVELS in "normal" adults in Western industrial societies, indicated by the curve, is compared with the range in adult animals and human infants and with the levels seen in FH patients. Levels in the shaded region of the chart are above the threshold associated with accelerated atherosclerosis; more than half of the adults have LDL levels above the threshold. The LDL level is inversely associated with the number of LDL receptors (*color*).

level rises in a population. As long ago as 1958 Ancel Keys of the University of Minnesota Medical School studied populations, in seven countries, in which the mean total cholesterol level varied from a high of 265 milligrams per deciliter to a low of 160. (He did not measure LDL cholesterol specifically, but because the level of lipoproteins other than LDL does not vary much, one can assume that the variations in total cholesterol reflected differences in LDL level.) Keys recorded the cholesterol level of 12,763 age-matched men in the seven countries, and 10 years later he determined which of the men had had a heart attack.

Two variables were found to correlate strongly with cholesterol level: the incidence of coronary atherosclerosis (as measured by fatal heart attacks) and the dietary intake of animal fats. In two villages (in Japan and Yugoslavia) where the mean total cholesterol level was 160 the incidence of fatal heart attacks was less than five per 1,000 men per 10 years. In eastern Finland, where the mean total cholesterol level was 265, the incidence of fatal heart attacks was 14 times as high. In populations with intermediate cholesterol levels (as in the U.S.) the incidence fell between the two extremes.

The correlation between cholesterol level and dietary intake of animal fats was even stronger than the correlation between cholesterol and atheroscle-

rosis. Populations consuming small amounts of animal fats (as in Japan and Yugoslavia) had low cholesterol levels. Populations with a high intake of such fats (as in eastern Finland) had high levels. Subsequent studies of many different populations have confirmed Keys's findings: high LDL levels are the rule in populations that consume a large part of their calories as fats from meat and dairy products.

The LDL-receptor hypothesis provides a likely explanation of the epidemiologic data. A high average intake of cholesterol makes cholesterol accumulate in liver cells. The accumulation seems to be accentuated by ingestion of animal fats rich in saturated fatty acids. Even a modest accumulation of cholesterol in the liver would partially suppress the manufacture of LDL receptors. This could lead to an increase in the average LDL level that would be detectable in an entire population.

Animal experiments by our group and by Mahley and Innerarity support the hypothesis that a high-fat diet reduces LDL receptors in the liver. In baboons, rabbits and dogs maintained on low-fat diets the number of LDL receptors is high and the animals degrade injected LDL rapidly; their LDL level is much lower than it is in human beings. When rabbits and dogs are fed diets high in cholesterol, their manufacture of receptors in the liver is suppressed by as much as 90 percent, and the result is a buildup of both IDL and LDL in the bloodstream. At birth human infants have LDL concentrations similar to those of other animal species; apparently newborn human beings make a large number of LDL receptors. During the childhood and early-adult years in industrialized societies, however, the LDL levels rises three- or fourfold. Studies in adults injected with LDL suggest that the increase is attributable to a decrease in the number of receptors with age.

The causes of the acquired receptor deficiency in human beings are not all known. The high dietary intake of animal fats seems to be an important factor, but it is not the only one: even in people raised on diets extremely low in fats the LDL level tends to be higher than it is in other species. Such hormones as estradiol and thyroid hormone are known to stimulate the manufacture of LDL receptors in the liver, and it is possible that subtle abnormalities in these and other hormones contribute to the age-related decrease in receptors.

The concentration of LDL eventually attained in most middle-aged adults in the U.S. and in similar societies is associated by epidemiological data with accelerated atherosclerosis. Experiments with cultured cells show why. The receptors bind LDL optimally when it is present in the blood at a concentration below 50 milligrams per deciliter. The receptors in animals and in humans (judging by the LDL level in human infants) have apparently been selected by evolution to function at just such levels. Yet in Western industrial countries the average "normal" LDL level in adults is about 125 milligrams per deciliter, considerably above the concentration at which receptors bind LDL most efficiently.

One finding that is consistent with the LDL-receptor hypothesis has been reported by William R. Hazzard of the Johns Hopkins Hospital and his colleagues. They showed that ingestion by adults of a high-cholesterol diet (including three egg yolks per day) does lead to a decrease in the number of LDL receptors, which they measured directly in circulating lymphocytes. A definitive test of the hypothesis will, however, require a comprehensive and well-controlled study of the rate of metabolism of injected VLDL and LDL in members of populations with low-fat and high-fat diets and with varying LDL levels. That has not yet been done systematically.

If the LDL-receptor hypothesis is correct, the human receptor system is designed to function in the presence of an exceedingly low LDL level. The kind of diet necessary to maintain such a level would be markedly different from the customary diet in Western industrial countries (and much more stringent than moderate low-cholesterol diets of the kind recommended by the American Heart Association). It would call for total elimination of dairy products as well as eggs, and for a severely limited intake of meats and other sources of saturated fats.

We believe such an extreme dietary change is not warranted for the entire population. There are several reasons. First, such a radical change in diet would have severe economic and social consequences. Second, it might well expose the population to other diseases now prevented by a moderate intake of fats. Third, experience shows most Americans will not adhere voluntarily to an extreme low-fat diet. Fourth, and most compelling, people vary genetically. Among those who consume the current high-fat diet of Western industrial societies, only 50 percent will die of atherosclerosis; the other 50 percent are resistant to the disease.

Some individuals resist atherosclerosis because their LDL level does not rise dangerously even though they consume a high-fat diet; they may inherit genes that somehow circumvent the usual feedback system and maintain receptor manufacture at an adequate level. Barbara V. Howard of the National Institutes of Health Clinical Research Center in Phoenix has shown, for example, that Indians of the Pima tribe have relatively large numbers of LDL receptors, and maintain low LDL levels, in spite of a high-fat diet. In other individuals the arteries apparently resist the damaging effects of elevated LDL. For example, 20 percent of men with heterozygous FH do not have a heart attach before the age of 60 even though their blood LDL is very high.

Given these reasons for constraint, what can be done to prevent accelerated atherosclerosis? One approach is to individualize dietary recommendations. A diet moderately low in animal fats would seem to be prudent for most people. The diet proposed by the American Heart Association, for example, would reduce blood cholesterol levels by as much as 15 percent and should somewhat lessen the incidence of heart attacks. On the other hand, people who have a strong family history of heart attacks or strokes, and who may therefore be particularly susceptible to the damaging effects of LDL, might well be encouraged to follow a diet extremely low in cholesterol and saturated fats—even if their LDL level is near the mean "normal" level. One can hope additional research will identify factors that either sensitize people to the ill effects of LDL or protect them from those effects.

Finally, therapy with drugs that increase the number of LDL receptors may turn out to be appro-priate for at least some people who do not have FH but in whom the number of receptors is reduced by diet or other factors. if it is shown that these drugs do prevent diet-induced suppression of receptors and if the drugs can be shown to be safe for long-term use, it may one day be possible for many people to have their steak and live to enjoy it too.

The Authors

MARC CANTIN and JACQUES GENEST ("The Heart as an Endocrine Gland") work at the Clinical Research Institute of Montreal, where Cantin is director of the laboratory of pathobiology and of a multidisciplinary research group on hypertension and where Genest is a consultant. They are both professors at McGill University and the University of Montreal. Cantin has a B.A. (1953) and an M.D. (1958) from Laval University and a Ph.D. (1962) from the University of Montreal. From 1962 to 1965 he was at the University of Chicago and joined the staff at Montreal in 1965 and McGill in 1983. Genest's B.A. (1937) and his M.D. (1942) are from Montreal. He did postgraduate work at the Hôtel-Dieu Hospital in Montreal, at Johns Hopkins, at Harvard University and at the Rockefeller Institute. He has practiced at Hôtel-Dieu since 1952. He joined Montreal in 1965 and McGill in 1970.

KERSTIN UVNÄS-MOBERG ("The Gastrointestinal Tract in Growth and Reproduction") is senior lecturer in the department of pharmacology at the Karolinska Institute in Stockholm. She received her M.D. from the institute in 1970, her Ph.D. in 1976.

DAVID W. GOLDE and JUDITH C. GASSON ("Hormones that Stimulate the Growth of Blood Cells") are colleagues at the University of California at Los Angeles School of Medicine who collaborate in research on the colony-stimulating factors. Golde, professor of medicine, serves as chief of the Division of Hematology-Oncology, director of the UCLA AIDS Center and director of the UCLA General Clinical Research Center. Gasson, assistant professor of medicine, is associate director of the division Golde heads.

MICHAEL J. BERRIDGE ("The Molecular Basis of Communication within the Cell") is senior principal scientific officer in the unit of insect neurophysiology and pharmacology at the University of Cambridge. He received a B.Sc. from the University College of Rhodesia and Nyasaland in 1960. His Ph.D. was awarded by the University of Cambridge in 1965. He spent a year at the University of Virginia and three years at Case Western Reserve University and returned to Cambridge in 1969.

WAI YIU CHEUNG ("Calmodulin") is member in biochemistry at St. Jude Children's Research Hospital in Memphis, Tenn. He was graduated from the National Tsung Hsing University in Taiwan with a B.S. in 1956. He obtained his M.S. at the University of Vermont in 1960 and his Ph.D. from Cornell University in 1964. From 1964 to 1967 he was research fellow at the University of Pennsylvania. In 1967 he moved to St. Jude. Since 1976 he has also been professor of biochemistry at the University of Tennessee Center for the Health Sciences in Memphis.

HOWARD RASMUSSEN ("The Cycling of Calcium as an Intracellular Messenger") is professor of internal medicine and of cellular and molecular physiology at Yale University School of Medicine. He has had a lifelong interest in the human metabolism of calcium and has published hundreds of articles and a book on the calcium-messenger system. Rasmussen earned an M.D. at Harvard University in 1952 and a Ph.D. at Rockefeller University in 1959. He worked at the University of Wisconsin and the University of Pennsylvania School of Medicine before moving to Yale in 1976.

LUBERT STRYER ("The Molecules of Visual Excitation") is Winzer Professor of Cell Biology at the Stanford University School of Medicine. He earned a B.S. at the University of Chicago in 1957 and an M.D. at the Harvard Medical School in 1961. Between 1961 and 1964 he held a Helen Hay Whitney research fellowship, spending part of his tenure as a visiting investigator at the Laboratory of Molecular Biology in Cambridge, England. He taught biochemistry at Stanford between 1963 and 1969, then was professor of molecular biophysics and biochemistry at Yale University. He returned to Stanford in 1976.

ALICE DAUTRY-VARSAT and HARVEY F. LODISH ("How Receptors Bring Proteins and Particles into Cells") are molecular biologists who worked together at the Massachusetts Institute of Technology.

Dautry-Varsat received a master's degree in solid-state physics at the University of Paris and a second master's degree in molecular biology at the State University of New York at Stony Brook. She earned her doctorate at the Pasteur Institute in Paris, then worked at the Medical Research Council Laboratory of Molecular Biology in England and at MIT before joining the Pasteur Institute. Lodish is professor of biology at MIT. He was graduated from Kenyon College and received his Ph.D. from Rockefeller University. He joined MIT in 1968, and in 1984 moved his laboratory to the Whitehead Institute for Biomedical Research on the MIT campus.

EDWARD RUBENSTEIN ("Diseases Caused by Impaired Communication among Cells") is associate dean of postgraduate medical education and professor of clinical medicine at the Stanford University School of Medicine, where he has been a member of the faculty for the past 25 years. He is also editor-in-chief of *Scientific American Medicine*. He received his M.D. from the University of Cincinnati College of Medicine in 1947.

MICHAEL B. A. OLDSTONE ("Viral Alteration of Cell Function") is a member of the departments of immunology and neuropharmacology at Scripps Clinic and Research Foundation in La Jolla, Calif., where he heads a laboratory of viral immunobiology. Oldstone received a B.S. from the University of Alabama in 1954 and his M.D. from the University of Maryland School of Medicine in 1961. He joined Scripps in 1969 and since 1972 has served as adjunct professor of pathology and neuroscience at the University of California, San Diego.

ABNER LOUIS NOTKINS ("The Causes of Diabetes") is chief of the Laboratory of Oral Medicine at the National Institute of Dental Research. He did his undergraduate work at Yale College, received his M.D. from the New York University School of Medicine in 1958 and did his internship and residency in internal medicine at the Johns Hopkins Hospital. He joined the National Institutes of Health as a research associate at the National Cancer Institute and moved to the Institute of Dental Research in 1973.

ANTHONY CERAMI, HELEN VLASSARA and MICHAEL BROWNLEE ("Glucose and Aging") share an interest in diabetes and in the aging process. Cerami is R. Gwin Follis-Chevron professor as well as dean of graduate and postgraduate studies and head of the laboratory of medical biochemistry at Rockefeller University. He was educated at Rutgers University and at Rockefeller, where he was awarded his doctorate in 1967. Vlassara is assistant professor of medical biochemistry at Rockefeller University. She earned her medical degree at the University of Athens School of Medicine and then came to the U.S. for an internship and residency at the Columbia University College of Physicians and Surgeons. Brownlee, who is codirector of the Diabetes Research Center at Albert Einstein College of Medicine, holds an M.D. from the Duke University School of Medicine. He did his internship and residency at the Stanford University School of Medicine and in 1975 became a research fellow in biological chemistry at the Harvard Medical School. After a fellowship at the Joslin Diabetes Center and the New England Deaconess Hospital and a stint as an instructor at Harvard, he went to Rockefeller and then to Albert Einstein.

MICHAEL S. BROWN and JOSEPH L. GOLDSTEIN ("How LDL Receptors Influence Cholesterol and Atherosclerosis") are professors of medicine and genetics at the University of Texas Health Science Center at Dallas, where Brown is also director of the Center for Genetic Disease and Goldstein is chairman of the department of molecular genetics. Brown received his M.D. from the University of Pennsylvania School of Medicine in 1966. He did his internship and residency at the Massachusetts General Hospital in Boston, then began several years of research at the National Institute of Arthritis and Metabolic Diseases and the National Heart Institute's laboratory of biochemistry. In 1971 Brown joined the University of Texas faculty. Goldstein is a 1966 graduate of the University of Texas Southwestern Medical School. He completed his medical training at the Massachusetts General Hospital and spent four years doing research on molecular genetics at the National Heart Institute and the University of Washington School of Medicine in Seattle. He moved to the University of Texas in 1972.

Bibliographies

1. The Heart as an Endocrine Gland

Marie, J. P., H. Guillemot and P. Y. Hatt. 1976. Le degré de granulation des cardiocytes auriculaires. Étude planimétrique au cours de différents apports d'eau et de sodium chez le rat. *Pathologie Biologie* 24 (October): 549–554.

Wardener, H. E. de, and G. A. MacGregor. 1983. The natriuretic hormone and its possible relationship to hypertension. In *Hypertension: Physiopathology and treatment*, eds. J. Genest, O. Kuchel, P. Hamet and M. Cantin. McGraw-Hill Book Company.

Burnett, John C., Jr., Joey P. Granger and Terry J. Opgenorth. 1984. Effects of synthetic atrial natriuretic factor on renal function and renin release. *American Journal of Physiology* 247: *Renal Fluid and Electrolyte Physiology* 16 (November): F863–F866.

Cantin, M., and J. Genest. 1985. The heart and the atrial natriuretic factor. *Endocrine Reviews* 6 (Spring): 107–127.

2. The Gastrointestinal Tract in Growth and Reproduction

Uvnäs-Moberg, Kerstin. 1987. Gastrointestinal hormones and pathophysiology of functional gastrointestinal disorders. *Scandinavian Journal of Gastroenterology* 22:138–146.

Numan, Michael. 1988. Maternal behavior. In *The physiology of reproduction*, eds. E. Knobil et al. Raven Press.

Uvnäs-Moberg, Kerstin. 1989. Neuroendocrine regulation of hunger and satiety. In *Obesity in Europe*, vol. 1, eds. P. Björntorp and S. Rössner. John Libbey & Company, Ltd.

Uvnäs-Moberg, Kerstin, and J. Winberg. 1981. Role for sensory stimulation in energy economy of mother and infant with particular regard to the gastrointestinal endocrine system. In *Textbook of gastroenterology and nutrition in infancy*, 2nd ed., ed. E. Lebenthal. Raven Press.

3. Hormones that Stimulate the Growth of Blood Cells

Golde, David W., and Takaku Fumimaro, eds. 1985. *Hematopoietic stem cells*. Marcel Dekker, Inc.

Metcalf, Donald. 1985. The granulocyte-macrophage colony-stimulating factors. *Science* 229 (July 5); 16–22.

Clark, Steven C., and Robert Kamen. 1987. The human hematopoietic colony-stimulating factors. *Science* 236 (June 5); 1229–1237.

Golde, David W., and J. C. Gasson. 1988. Myeloid growth factors. In *Inflammation: Basic principles and clinical correlates*, eds. J. I. Gallin, I. M. Goldstein and R. Synderman. Raven Press.

4. The Molecular Basis of Communication within the Cell

Cohen, Philip. 1982. The role of protein phosphorylation in neural and hormonal control of cellular activity. *Nature* 296 (April 15): 613–620.

Bishop, J. Michael. 1983. Cellular oncogenes and retroviruses. *Annual Review of Biochemistry* 52:301–354.

Gilman, Alfred G. 1984. G proteins and dual control of adenylate cyclase. *Cell* 36 (March): 577–579.

Nishizuka, Yasutomi. 1984. The role of protein kinase C in cell surface signal transduction and tumor promotion. *Nature* 308 (April 19): 693–698.

Berridge, M. J., and R. F. Irvine. 1984. Inositol triphosphate, a novel second messenger in cellular signal transduction. *Nature* 312 (November 22): 315–321.

5. Calmodulin

Wang, Jerry J., and David Morton Waisman. 1979. Calmodulin and its role in the second-messenger system. *Current Topics in Cellular Regulation* 15:47–107.

Cheung, Wai Yiu, ed. 1980. *Calcium and cell function*, vol I. Academic Press.

Klee, C. B., T. H. Crouch and P. G. Richman. 1980. Calmodulin. *Annual Review of Biochemistry* 49:489–515.

Watterson, D. Martin, and Frank F. Vincenzi, eds. 1980. Calmodulin and cell function. *Annals of the New York Academy of Sciences* 356:1–443.

Cheung, Wai Yiu. 1980. Calmodulin plays a pivotal role in cellular regulation. *Science* 207 (January 4): 19–27.

———. 1981. Discovery and recognition of calmodulin: A personal account. *Journal of Cyclic Nucleotide Research* 7:71–84.

6. The Cycling of Calcium as an Intracellular Messenger

Rasmussen, Howard, 1981. *Calcium and cAMP as synarchic messengers.* John Wiley & Sons, Inc.

Campbell, Anthony K. 1983. *Intracellular calcium: Its universal role as regulator.* John Wiley & Sons, Inc.

Rasmussen, Howard. 1986. The calcium messenger system. *New England Journal of Medicine* 314 (April 24): 1094–1101; and 314 (May 1): 1164–1170.

Alkon, Daniel L., and Howard Rasmussen. 1988. A spatial-temporal model of cell activation. *Science* 239 (February 26): 998–1005.

Adam, Leonard P., Joe R. Haeberle and David R. Hathaway. 1989. Phosphorylation of caldesmon in arterial smooth muscle. *Journal of Biological Chemistry* 264 (May 5): 7698–7703.

7. The Molecules of Visual Excitation

Liebman, P. A., and E. N. Pugh, Jr. 1981. Control of rod disk membrane phosphodiesterase and a model for visual transduction. *Current Topics in Membranes and Transport* 15:157–170.

Fung, Bernard K.-K., James B. Hurley and Lubert Stryer. 1981. Flow of information in the light-triggered cyclic nucleotide cascade of vision. *Proceedings of the National Academy of Sciences* 78 (January): 152–156.

Vuong, T. M., M. Chabre and Lubert Stryer. 1981. Millisecond activation of transducin in the cyclic nucleotide cascade of vision. *Nature* 311 (October 18): 659–661.

Stryer, Lubert. 1986. Cyclic GMP cascade of vision. *Annual Review of Neuroscience* 9:87–119.

8. How Receptors Bring Proteins and Particles into Cells

Geuze, Hans J., Jan Willem Slot, Ger J. A. M. Strous, Harvey F. Lodish and Alan L. Schwartz. 1983. Intracellular site of asialoglycoprotein receptor-ligand uncoupling: Double-label immunoelectron microscopy during receptor-mediated endocytosis. *Cell* 32 (January): 277–287.

Dautry-Varsat, Alice, Aaron Ciechanover and Harvey F. Lodish. 1983. *p*H and the recycling of transferrin during receptor-mediated endocytosis. *Proceedings of the National Academy of Sciences* 80 (April): 2258–2262.

9. Diseases Caused by Impaired Communication among Cells

Goldstein, Avram, Lewis Aronow and Sumner M. Kalman. 1974. *Principles of drug action.* John Wiley & Sons.

Brown, Michael S., and Joseph L. Goldstein. 1976. Familial hypercholesterolemia: A genetic defect in the low-density lipoprotein receptor. *New England Journal of Medicine* 294 (June 17): 1386–1390.

Bar, Robert S., and Jesse Roth. 1977. Insulin receptor status in disease states of man. *Archives of Internal Medicine* 137 (April): 474–481.

Barchas, Jack D., Huda Akil, Glen R. Elliott, R. Bruce Holman and Stanley J. Watson. 1978. Behavioral neurochemistry: Neuroregulators and behavioral states. *Science* 200 (May 26): 964–973.

Mehdi, S. Qasim, and Joseph P. Kriss. 1978. Preparation of radiolabeled thyroid-stimulating immunoglobulins (TSI) by recombing TSI heavy chains with ^{125}I-labeled light chains: Direct evidence that the product binds to the membrane thyrotropin receptor and stimulates adenylate cyclase. *Endocrinology* 103 (July): 296–301.

Goldfine, I. D., A. L. Jones, G. T. Hradek, K. Y. Wong and J. S. Mooney. 1978. Entry of insulin into human cultured lymphocytes: Electron microscope autoradiograph analysis. *Science* 202 (November 17): 760–762.

10. Viral Alteration of Cell Function

Oldstone, Michael B. A. 1984. Virus can alter cell function without causing cell pathology. In *Concepts in viral pathogenesis*, eds. Abner Louis Notkins and Michael B. A. Oldstone, Springer-Verlag.

Oldstone, Michael B. A., et al. 1986. Cytoimmunotherapy for persistent virus infection reveals a unique clearance pattern from the central nervous system. *Nature* 321 (May 15): 239–243.

McChesney, M. B., and M. B. A. Oldstone. 1987 Viruses perturb lymphochyte functions. *Annual Review of Immunology* 5:279–304.

Oldstone, Michael B. A. 1989. Viral persistence. *Cell* 56 (February 24): 517–520.

Klavinskis, Linda S., and Michael B. A. Oldstone. 1989. Lymphocytic chriomeningitis virus selectively alters differentiation but not housekeeping functions. *Virology* 168:232–235.

11. The Causes of Diabetes

U.S. Department of Health, Education, and Welfare. 1976. *Report on the National Commission on Diabetes to the Congress of the United States.* Government Printing Office.

Eastman, Richard C., and Jeffrey S. Flier. 1977. Receptors for peptide hormones: New insights into the pathopsychology of disease states in man. *Annals of Internal Medicine* 86 (February): 205–219.

Cudworth, A. G. 1978. Type I diabetes mellitus. *Diabetologia* 14 (May): 281–291.

Yoon, Ji-Won, Marshall Austin, Takashi Onodera and Abner Louis Notkins. 1979. Virus-induced diabetes mellitus: Isolation of a virus from the pancreas of a child with diabetic ketoacidosis. *New England Journal of Medicine* 300 (May 24): 1173–1179.

12. Glucose and Aging

Monnier, Vincent M., Robert R. Kohn and Anthony Cerami. 1984. Accelerated age-related browning of human collagen in diabetes mellitus. *Proceedings of the National Academy of Sciences* 81 (January): 583–587.

Bucala, Richard, Peter Model and Anthony Cerami. 1984. Modification of DNA by reducing sugars: A possible mechanism for nucleic acid aging and age-related dysfunction in gene expression. *Proceedings of the National Academy of Sciences* 81 (January): 105–109.

Vlassara, Helen, Michael Brownlee and Anthony Cerami. 1985. High-affinity receptor-mediated uptake and degradation of glucose-modified proteins: A potential mechanism for the removal of senescent macromolecules. *Proceedings of the National Academy of Sciences* 82 (September): 5588–5592.

Brownlee, Michael, Helen Vlassara, A. Kooney, P. Ulrich and A. Cerami. 1986. Aminguanidine prevents diabetes-induced arterial wall protein cross-linking. *Science* 232 (June 27): 1629–1632.

13. How LDL Receptors Influence Cholesterol and Atherosclerosis

Goldstein, Joseph L., and Michael S. Brown. 1983. Familial hypercholesterolemia. In *The Metabolic basis of inherited disease*, eds. John B. Stanbury, James B. Wyngaarden, Donald S. Frederickson, Joseph L. Goldstein and Michael S. Brown. McGraw-Hill Book Company.

Mahley, Robert W., and Thomas L. Innerarity. 1983. Lipoprotein receptors and cholesterol homeostasis. *Biochimica et Biophysica* Acta 737 (May 24): 197–222.

Bilheimer, David W., Scott M. Grundy, Michael S. Brown and Joseph L. Goldstein. 1983. Mevinolin and colestipol stimulate receptor-mediated clearance of low-density lipoprotein from plasma in familial hypercholesterolemia heterozygotes. *Proceedings of the National Academy of Sciences* 80 (July): 4124–4128.

Goldstein, Joseph L., Toru Kita and Michael S. Brown. 1983. Defective lipoprotein receptors and atherosclerosis: Lessons from an animal counterpart of familial hypercholesterolemia. *New England Journal of Medicine* 309 (August 4): 228–296.

Sources of the Photographs

Marc Cantin: Figure 1.1

National Art Museums of Sweden: Figure 2.7

David W. Golde: Figures 3.5, 3.6 and 3.9

Tripos Associates: Figure 4.1

Jeffrey F. Harper and Alton L. Steiner: Figures 5.7 and 5.11

John G. Wood: Figure 5.9

John R. Dedman, B. R. Brinkley and Anthony F. Means: Figure 5.10

M. M. Perry and A. B. Gilbert, Agricultural Research Council Poultry Research Center, Edinburgh: Figure 8.2

J. Geuze, State University of Utrecht: Figure 8.5

Frederick R. Maxfield, New York University Medical Center: Figure 8.6

Richard G. W. Anderson, University of Texas Health Science Center at Dallas: Figure 9.7

I. D. Goldfine, University of California at San Francisco: Figure 9.8

Michael B. A. Oldstone: Figures 10.2 and 10.6

UPI/Bettman Newsphotos: Figure 10.4

A. Bennett Jenson and Kozaburo Hayashi, National Institute of Dental Research: Figure 11.2

Joseph R. Williamson, Washington University School of Medicine: Figure 11.5

Takashi Onodera, National Institute of Dental Research: Figure 11.6

A. Bennett Jenson: Figure 11.11

David M. Phillips, The Population Council: Figure 12.2

Christian McBride and David E. Birk, University of Medicine and Dentistry of New Jersey, Robert Wood Johnson Medical School, Piscataway: Figure 12.4

L. Maximilian Buja, University of Texas Health Science Center at Dallas: Figure 13.1

INDEX

Page numbers in *italics* indicate illustrations.